Lecture Notes in Statistics

Edited by J. Berger, S. Fienberg, J. Gani,
K. Krickeberg, I. Olkin, and B. Singer

71

Eduardo M.R.A. Engel

A Road to Randomness in Physical Systems

Springer-Verlag
Berlin Heidelberg New York London Paris
Tokyo Hong Kong Barcelona Budapest

Author

Eduardo M. R. A. Engel
Department of Economics
Massachusetts Institute of Technology
Cambridge, MA 02139, USA
and
Departamento de Ingeniería Industrial
Universidad de Chile
República 701, Santiago, Chile

Mathematical Subject Classification: 60A99, 60E05, 60F99, 58F07, 58F11

ISBN 0-387-97740-6 Springer-Verlag New York Berlin Heidelberg
ISBN 3-540-97740-6 Springer-Verlag Berlin Heidelberg New York

This work is subject to copyright. All rights are reserved, whether the whole or part of the material is concerned, specifically the rights of translation, reprinting, re-use of illustrations, recitation, broadcasting, reproduction on microfilms or in any other way, and storage in data banks. Duplication of this publication or parts thereof is permitted only under the provisions of the German Copyright Law of September 9, 1965, in its current version, and permission for use must always be obtained from Springer-Verlag. Violations are liable for prosecution under the German Copyright Law.

© Springer-Verlag Berlin Heidelberg 1992
Printed in the United States of America

Typesetting: Camera ready by author

47/3140-543210 – Printed on acid-free paper

For my parents

Preface

There are many ways of introducing the concept of probability in classical, i.e. deterministic, physics. This work is concerned with one approach, known as *"the method of arbitrary functions."* It was put forward by Poincaré in 1896 and developed by Hopf in the 1930's. The idea is the following. There is always some uncertainty in our knowledge of both the initial conditions and the values of the physical constants that characterize the evolution of a physical system. A probability density may be used to describe this uncertainty. For many physical systems, dependence on the initial density washes away with time. In these cases, the system's position eventually converges to the same random variable, no matter what density is used to describe initial uncertainty.

Hopf's results for the method of arbitrary functions are derived and extended in a unified fashion in these lecture notes. They include his work on dissipative systems subject to weak frictional forces. Most prominent among the problems he considers is his carnival wheel example, which is the first case where a probability distribution cannot be guessed from symmetry or other plausibility considerations, but has to be derived combining the actual physics with the method of arbitrary functions. Examples due to other authors, such as Poincaré's law of small planets, Borel's billiards problem and Keller's coin tossing analysis are also studied using this framework. Finally, many new applications are presented. They include bouncing balls, physical systems described by small oscillations (such as a simple pendulum and a coupled harmonic oscillator) and integrable systems (such as a heavy symmetric top).

This work shows that, from a mathematical point of view, most applications of the method of arbitrary functions follow from the fact that the fractional part of the product of a real number t and an absolutely continuous random vector X converges to a distribution uniform on the unit hypercube as t tends to infinity. This motivates the study of the speed at which convergence takes place in order to determine the practical relevance of the method of arbitrary functions for specific examples. New results on convergence, and tractable upper bounds for the rate of convergence are derived and applied.

Chapter 1 gives a more detailed overview of the contents of this work. Mathematical preliminaries are considered in Chap. 2. Mathematical results are presented in Sect. 3.1 (for the one dimensional case) and Sect. 4.1 (for higher dimensions). These results are applied to various examples in Sects. 3.2 and 4.2. The sections with applications may be read independently from the ones containing the mathematical results.

Hopf's contribution to the method of arbitrary functions is studied in detail in Chap. 5. The concrete examples where he applied this method – including his carnival wheel and Buffon needle examples – are considered in Sects. 5.1, 5.2 and 5.3. The concept of *statistical regularity* which he introduced to formalize the idea of the unpredictability of a dynamical system – in the sense of the method of arbitrary functions – and its close connection to ergodic theory are discussed in Sects. 5.4 and 5.6. The relation he established between the physical and statistical independence of variables describing a dynamical system is discussed in Section 5.5. This chapter concludes by extending the concept of statistical regularity to that of *partial* statistical regularity, thereby encompassing all examples considered in this work.

Chapter 6 studies the behavior of the fractional part of the random vectors $A^k X$ and $e^{tA} X$ when the integer k and the real number t tend to infinity, where A denotes an n by n matrix and X an n-dimensional absolutely continuous random vector. Necessary and sufficient conditions for the existence of a limit random variable that does not depend on the density associated with X are derived. They involve number theoretic properties of the generalized eigenvectors of the matrix A.

This work corresponds to the Ph. D. dissertation I wrote while studying Statistics at Stanford University. I thank Kevin Coakley, Bradley Efron, Joseph Keller, Joseph Marhoul, Iain Johnstone, David Siegmund and Charles Stein for insightful comments and suggestions. I am especially indebted to Persi Diaconis for suggesting this research topic and providing generous advice throughout the writing of this work.

Cambridge, Massachusetts, 1992   *Eduardo Engel*

Table of Contents

List of Figures

1. Introduction

There are many ways of introducing the concept of probability in classical, i.e. deterministic physics. This work is concerned with one approach, known as "the method of arbitrary functions." To illustrate it consider the following example:

1.1 The Simple Harmonic Oscillator

A block of mass m is attached to an ideal spring and free to move over a frictionless horizontal table (see Fig. 1.1). The spring is stretched a distance from its equilibrium position and released from rest.

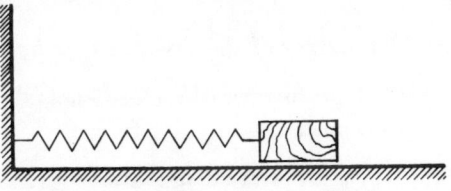

Fig. 1.1. Block and spring

If the initial deformation is not too large, Hooke's law applies. The force exerted on the spring when it is a distance x away from its equilibrium position is $-kx$, where k is a measure of the spring's stiffness called the *force constant*. Solving Newton's equation implies that the spring's displacement from its equilibrium position at time t is

$$x(t) = A\cos(\omega t), \tag{1.1}$$

where A denotes the spring's initial displacement and $\omega = \sqrt{\frac{k}{m}}$.

Equation (1.1) is a particular case of *simple harmonic motion*. There seems to be no other law that appears in so many different fields of physics: Maor (1972) lists 16 distinct phenomena governed by the simple harmonic law. They include examples in elasticity, gravitation, acoustics and hydrostatics.

The quantities ω and A in (1.1) are generally called the *angular frequency* and *amplitude* of the oscillator.

If the spring's constant and the block's mass and initial displacement are known exactly, its position at any instant can be determined from (1.1). However, in real life there is always some uncertainty in our knowledge of the spring's constant, k, the block's mass, m, and initial displacement, A. It is natural to think of them as random variables. The spring's displacement at time t, $x(t)$, then also is a random variable and its distribution depends on that of k, m and A. Just after releasing the block, dependence of $x(t)$ on k, m and A can be expected to be strong.

Does dependence decrease as time passes?

From (1.1) it is clear that the spring's maximum displacement is equal to its amplitude and that it takes this value an infinite number of times as time goes on. The probability that $x(t)$ takes large values depends on how large the amplitude is judged to be. Hence dependence of $x(t)$ on A does not wash away with time.

To see what happens with dependence on ω (or equivalently on m and k) consider

$$\frac{x(t)}{A} = \cos(\omega t) \qquad (1.2)$$

for large values of t. If the difference between times t_1 and t_2 is an integer multiple of 2π, the position of the spring at those instants is the same. Therefore (1.2) is equivalent to

$$\frac{x(t)}{A} = \cos\left((\omega t)(\bmod 2\pi)\right),$$

with $x(\bmod d)$ denoting the remainder of the division of x by d.

The following mathematical problem is therefore equivalent to determining if the dependence of the spring's position on its angular frequency washes away with time:

Given a random variable X, what is the distribution of $(tX)(\bmod 1)$ for large values of t?

As t grows, the distribution of tX spreads out more and more on the real line (see Fig. 1.2). When considering $(tX)(\bmod 1)$, this distribution is folded back onto the unit interval, and the resulting distribution is nearly uniform.

The variation distance between two random variables is the largest absolute difference of the probabilities both random variables assign to any given event. In Theorem 5.3 it is proved that $(tX)(\bmod 1)$ converges in the variation distance to a distribution uniform on the unit interval if and only if X has a density (with respect to Lebesgue measure). Furthermore, the limiting random variable is independent of almost any random variable that does not depend on t (Theorem 3.3).

Translating back to the elastic spring, consider the normalized displacement $x(t)/A$. Equation (1.1) implies that, as time passes, the block's normalized displacement converges to the cosine of a distribution uniform on $[0, 2\pi]$. The limiting random variable, S, belongs to the arcsine family. Its density is (see Fig. 1.3):

$$f(s) = \frac{1}{\pi\sqrt{1 - s^2}} \; ; \qquad -1 < s < 1.$$

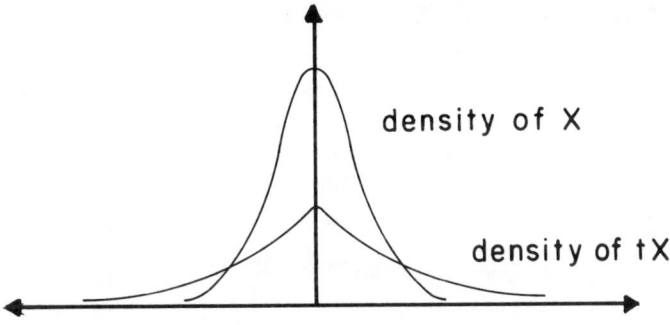

Fig. 1.2. Density of X and tX.

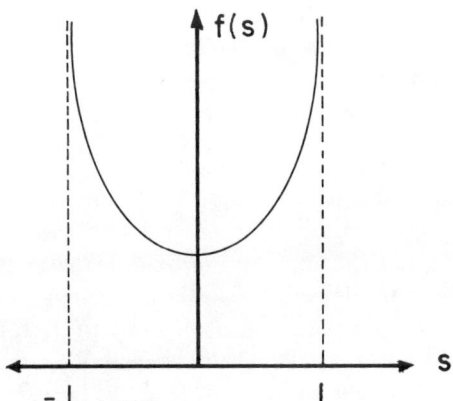

Fig. 1.3. Density of arcsine law

No matter what density is chosen to describe our uncertainty about the spring's constant and the block's mass, as time passes dependence on these random variables washes away. The limit random variable S is independent of the random variables describing m, k and A.

The block travels fastest when it passes through its equilibrium position and slows down to rest every time it approaches its maximum displacement. Therefore the probability of observing values of $x(t)$ near zero is smaller than that of observing values near its maximum displacement. This argument explains why the density of S takes its minimum value at zero and its maximum values at -1 and 1.

For the spring's displacement, $x(t) = A\cos(\omega t)$, the limit distribution is the product of the random variable describing its amplitude, A, and an independent arcsine law, S, that is, a scale mixture of arcsine laws.

1.2 Philosophical Interpretations

How does the concept of probability enter the world of classical, i.e. deterministic physics? Given a system's initial position and velocity, and the forces acting upon it, Newton's equations completely determine its motion at any instant of time. In principle there is no place for randomness. Yet people do speak about the probability of a roulette ball landing in a certain slot or the probability that a coin lands heads up.

One solution to this apparent contradiction – deterministic physics versus randomness – was put forward by Poincaré in 1896 and developed by Hopf in the 1930s. It is known as the *"method of arbitrary functions."* The idea is the following. There is always some uncertainty in our knowledge of initial conditions. A probability density may be used to describe this uncertainty. For many physical systems, dependence on the initial density washes away with time. In these cases, the system's position eventually converges to the same random variable, no matter what density is used to describe initial conditions. This explains the name of *"method of arbitrary functions."* The phenomena considered are such that the conclusions drawn about their outcome are valid for an *arbitrary* density describing initial conditions.

How should the density on initial conditions be interpreted? Poincaré (1905) and Hopf (1934, 1936, 1937) consider it a smooth approximation to the corresponding empirical distribution. A subjectivist view is due to Savage (1973). In the case of the oscillating spring, it implies that any two individuals with prior beliefs about the oscillator's angular frequency modeled by a probability density are forced to agree on the distribution of the block's normalized displacement, $x(t)/A$, as time passes. This is quite different from the usual argument for multisubjective agreement – *"the data swamp the prior"*– since it applies to a single event. Agreement is reached without the need of any data: *the physics swamps the prior.*

The results developed in this work may be interpreted both from a frequentist and a subjective point of view, the author has taken no position when discussing concrete applications and both approaches are considered.

Von Plato (1983) draws attention to the importance of the method of arbitrary functions for the foundations of probability, this paper also includes a useful survey of this topic.

1.3 Coupled Harmonic Oscillators

A simple pendulum consists of a small weight, suspended by a light inextensible cord. When pulled to one side of its equilibrium position and released, the pendulum swings in a vertical plane under the influence of gravity.

Consider a pair of coupled harmonic oscillators, that is, two simple pendulums whose bobs are connected by a spring, as indicated in Fig.1.4. For simplicity assume both bobs have the same mass, m, and both cords the same length, l. Define $k = mg/l$, where g

denotes acceleration due to gravity. Let k' denote the spring's constant and $x_1(t)$ and $x_2(t)$ the weights' displacement from their equilibrium position at time t.

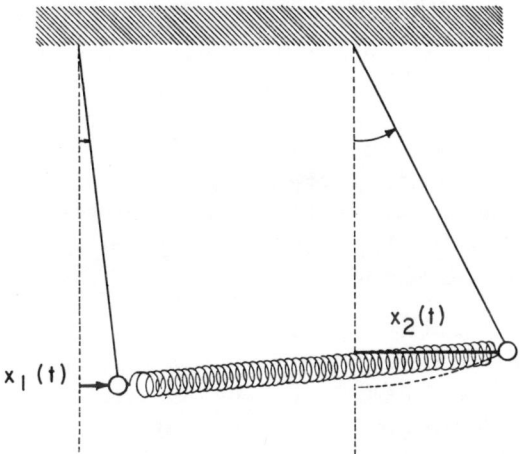

Fig. 1.4. Coupled harmonic oscillators

If initial displacements are small and both pendulums start from rest, Newton's equations lead to the following (approximate) solution (see Sect. 2.2 of Chap. 4 for their derivation):

$$x_1(t) = A \cos(\omega t) + B \cos(\omega' t) , \qquad x_2(t) = A \cos(\omega t) - B \cos(\omega' t),$$

with

$$A = \frac{x_1(0) + x_2(0)}{2} , \qquad B = \frac{x_1(0) - x_2(0)}{2},$$

and

$$\omega = \sqrt{\frac{k}{m}} , \qquad \omega' = \sqrt{\frac{k + 2k'}{m}}.$$

Assume the values of the physical constants k, k' and m and those of the initial displacements, $x_1(0)$ and $x_2(0)$, are not known exactly but described by a five dimensional random vector.

The random vector $(x_1(t), x_2(t))$ depends on time through the random variables $(\omega t)(\mathrm{mod}\, 2\pi)$ and $(\omega' t)(\mathrm{mod}\, 2\pi)$. This motivates considering the following mathematical problem:

Given an n dimensional random vector $X = (X_1, \ldots, X_n)$, denote by $(tX)(\mathrm{mod}\, 1)$ the random vector whose i-th coordinate is equal to $(tX_i)(\mathrm{mod}\, 1)$; $i = 1, \ldots, n$.

What is the distribution of $(tX)(\mathrm{mod}\, 1)$ for large values of t?

Theorem 5.3 shows that $(tX)(\mathrm{mod}\, 1)$ converges in the variation distance to a distribution uniform on the unit hypercube, $[0, 1]^n$, if and only if X has a density. The limiting random vector is independent of almost any random vector that does not depend on time.

In the case of the coupled oscillator this means that $(x_1(t), x_2(t))$ converges to a random vector $L = (L_1, L_2)$ with $L_1 = AS_1 + BS_2$, $L_2 = AS_1 - BS_2$, where S_1 and S_2 are independent, identically distributed arcsine laws with common density

$$f(s) = \frac{1}{\pi\sqrt{1 - s^2}}; \qquad -1 < s < 1.$$

Furthermore, S_1 and S_2 are independent of A and B. That is, as time passes the position of the coupled harmonic oscillator converges to a sum of mixtures of independent arcsine laws. The mixing random variables are determined by those describing the initial displacement of the bobs. Dependence on the random variables describing the physical constants k, k' and m washes away.

The limit random variables L_1 and L_2 both have zero mean and variance equal to $\frac{1}{4}\mathrm{E}(x_1^2(0) + x_2^2(0))$. Their correlation is equal to $2\mathrm{E}x_1(0)x_2(0)/(\mathrm{E}(x_1^2(0) + x_2^2(0)))$. If $x_1(0)$ and $x_2(0)$ both have zero mean and the same variance, the correlation between L_1 and L_2 is equal to that between $x_1(0)$ and $x_2(0)$.

As discussed in Sect. 2.2 of Chap. 4, the motion of any conservative system in the neighborhood of a configuration of stable equilibrium exhibits a behavior of the type just described and may be analyzed in a similar way. The coordinates describing such a system's position converge to the product of a stochastic matrix M (that depends on the density describing initial conditions) and a vector of independent, identically distributed arcsine laws independent of M.

1.4 Mathematical Results

One of the aims of this work is to point out the importance of studying the behavior of the random vector $(tX)(\mathrm{mod}\, 1)$ for large values of t.

In Chap. 5 Hopf's results for the method of arbitrary functions are derived in a unified fashion using this approach. Poincaré's original applications may also be derived using this framework. Many new applications which rely on the same mathematical tools are presented in this work, such as the coupled harmonic oscillator described above.

For this reason, a detailed study of the behavior of the random vector $(tX)(\mathrm{mod}\, 1)$ as t tends to infinity is undertaken. The case where X is one dimensional is considered in Chap. 3. These results are extended to higher dimensions in Chap. 4.

A necessary and sufficient condition for weak-star convergence of the fractional part of tX, $(tX)(\mathrm{mod}\, 1)$, to a distribution uniform on $[0, 1]^n$, U_n, as t tends to infinity, is established in Theorem 4.2 (Borel, Hopf, Kallenberg). The limit random vector is independent of initial conditions (Theorem 4.4, Hopf). The random vector $(tX)(\mathrm{mod}\, 1)$ converges to U_n in the variation distance if and only if X has a density (Theorem 5.3).

The random vector $(tX)(\mathrm{mod}\, 1)$ can converge to U_n at very slow rates (Proposition 3.8). In Theorem 3.9 (Kemperman) it is shown that the one dimensional random variable $(tX)(\mathrm{mod}\, 1)$ converges to a distribution uniform on $[0, 1]$ at a rate at least linear in t^{-1}.

if the density of X has bounded variation. This is the case, in particular, if the density of X has an integrable derivative.

The concept of bounded variation is extended to higher dimensions (Sect. 2.1 of Chap. 4) leading to the notion of bounded mean-conditional variation. Upper bounds for the variation distance between $(tX)(\mathrm{mod}\,1)$ and U_n that are linear in t^{-1} are derived for random vectors satisfying this condition (Theorem 4.9). They apply, in particular, if the density of X has integrable partial derivatives.

Convergence can take place at faster rates than linear in t^{-1} if specific families of distributions are considered (Proposition 3.11 and 4.14). Among the well known one dimensional distributions, convergence is fastest (rate e^{-ct^2}) for normal random variables (Proposition 3.11). This led Aldous, Diaconis and Kemperman to conjecture that this was the best possible rate. A counterexample is given by constructing a family of random vectors for which the distribution of $(tX)(\mathrm{mod}\,1)$ is identically uniform once t passes a certain threshold (Theorem 3.14 and Proposition 4.15).

The random vector $(tX)(\mathrm{mod}\,1)$ may be viewed as the product of a diagonal matrix (with all its diagonal elements equal to t) and the random vector X. In this sense it corresponds to the "diagonal case." Chapter 6 studies the "non diagonal case," that is, given an n dimensional random vector X and a collection of n by n matrices, $\{A(\tau);\ \tau \in T \subset \mathbb{R}\}$, attention is focused on conditions under which $(A(\tau)X)(\mathrm{mod}\,1)$ converges to a distribution uniform on the unit hypercube, U_n, as τ tends to infinity.

If the absolute value of all eigenvalues of the matrices $A(\tau)$ tend to infinity, bounds on the variation distance between $(A(\tau)X)(\mathrm{mod}\,1)$ and U_n are derived in Theorem 6.2. They depend on the norm of the inverse of $A(\tau)$ and a measure of smoothness for the density of X.

In Theorems 6.10 and 6.14, necessary and sufficient conditions for weak-star convergence of $(A^k X)(\mathrm{mod}\,1)$ and $(e^{tA}X)(\mathrm{mod}\,1)$ to a distribution uniform on $[0,1]^n$ as the integer k and real number t tend to infinity, respectively, are derived. They involve number theoretic properties of the generalized eigenvectors of the matrix A.

1.5 Calculating Rates of Convergence

The fact that most limit theorems in this work include rates of convergence is used in many applications.

Consider, for example, a simple pendulum, that is, a small weight suspended from a light inextensible cord (see Fig.1.5).

Assume the pendulum starts from rest and its initial displacement, A, is small. Then its position at time t, $x(t)$, is approximately equal to

$$x(t) = A \cos\left(\sqrt{\frac{g}{l}}\, t\right),$$

where g denotes acceleration due to gravity and l the cord's length.

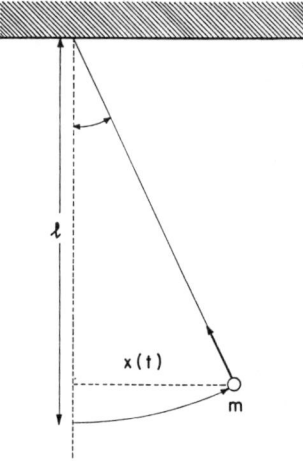

Fig. 1.5. Simple pendulum

As discussed at the beginning of this introduction, if uncertainty about the cord's length is described using a probability density, the pendulum's normalized displacement, $x(t)/A$, converges to an arcsine law, S.

After how much time is the variation distance between $x(t)/A$ and S less than, say, 0.05?

Calculations similar to those carried out in Sect. 2.1 of Chap. 3 show that if the distribution of l is a mixture of unimodal densities bounded by M with mode located at values not larger than x_0, then

$$ \mathrm{dv}\left(\frac{x(t)}{A}, S\right) \leq \frac{2x_0^{3/2} M + 3\sqrt{El}}{4t\sqrt{g}}, \tag{1.3} $$

where dv denotes variation distance, that is, the largest absolute difference between the probabilities assigned by both random variables to a given event, S an arcsine law and El the expected value of l.

Assume the cord is measured and its length lies somewhere between $11\frac{1}{2}$ and $12\frac{1}{2}$ inches. If l is uniform on this interval, (1.3) implies that the variation distance between $x(t)/A$ and U is less than 0.05 after 24 seconds. More generally, if l is assumed to be a mixture of unimodal densities with modes located at values smaller than $12\frac{1}{2}$ inches and having their size bounded by twice the maximum value of a distribution uniform on $[11\frac{1}{2}, 12\frac{1}{2}]$, (1.3) implies that the variation distance is less than 0.05 after 48 seconds. If A and l are assumed independent, these bounds also hold for the variation distance between the pendulum's displacement and the scale mixture of arcsine laws determined by A.

In this example it is not necessary to know the density describing initial conditions to obtain upper bounds for the variation distance between the random variable and its limit. All that is needed is an assessment of how smooth this density is. This is true in general.

1.6 Hopf's Approach

Though many authors extended Poincaré's work on the method of arbitrary functions, the most significant developments are due to Hopf in a series of papers published in the 1930s. Chapter 5 presents and extends his work in a unified fashion. Hopf's work on dissipative dynamical systems is studied in Sections 1, 2 and 3 of this chapter. As an example, consider a carnival wheel, that is, a wheel that spins on a central axis and is slowed down by friction and a rubber or leather pointer which bumps noisily against nails arranged on the wheel's rim (see Fig.1.6). The angular position of the wheel can be described by specifying the angular distance θ from the pointer to some mark on the wheel. Frictional forces slowing down the wheel depend on the wheel's position.

Fig. 1.6. Carnival wheel

Assume the wheel's initial position and velocity are described by a joint density. What is the distribution of the wheel's final position if it is given a large initial impulse? If no nails were present, symmetry considerations imply that if there exists a limiting distribution it has to be uniform on $[0, 2\pi]$. Yet no symmetry considerations are useful in this case because of the asymmetry introduced by the nails.

Hopf constructed a set of forces for which the distribution of the wheel's final position converges to a distribution that is *not* uniform. The limiting distribution has a very simple expression in terms of the particular force and does not depend on the joint density describing initial conditions. This seems to be the first example where a probability distribution cannot be guessed from symmetry or other plausibility considerations and has to be derived doing the actual physics.

This result can be used to show that if all slots on the carnival wheel have the same size but the number of nails per slot is not constant, the best bet is to play on the slots with more nails. Alternatively, if there are only nails at the beginning and end of each slot, but different slots have different arc lengths, the fair price is not inversely proportional to the arc length. Smaller slots have a larger probability of winning than their size would indicate.

1.7 Physical and Statistical Independence

The 1930's were years in which the axiomatization of the calculus of probability played an important role. From a mathematical point of view, what was new with respect to measure theory was the concept of independence. Quoting from Kolmogorov (1933, 1956 p.8ff):

"Historically the independence of experiments and random variables represents the very mathematical concept that has given the theory of probability its peculiar stamp. [...] In consequence, one of the most important problems in the natural sciences is – in addition to the well known one regarding the essence of the concept of probability itself – to make precise the premises which would make it possible to regard any given real events as independent. This question, however, is beyond the scope of this book."

Hopf was very interested in this problem. He mentioned the connection between the method of arbitrary functions and the concepts of physical and statistical independence in every article he wrote on this subject.

Physical independence of two variables is interpreted by Hopf as meaning that their initial distribution is described by a two dimensional density. Statistical independence of their final values should then be derived. If this program is successful, statistical independence follows from physical independence and need not be assumed.

Consider, for example, a dynamical system whose position at time t, $\phi(t)$, is determined by n coordinates and satisfies the following form of Newton's equation:

$$\phi''(t) = \mu F\big(\phi'(t)\big),$$

where F is a function of only velocity, μ a parameter that is interpreted as a friction coefficient, and $\phi'(t)$ and $\phi''(t)$ denote the system's velocity and acceleration vectors at time t. Assume the system eventually comes to rest (i.e. is dissipative) and that F is a continuous and positive with $F(0) = 0$. Hopf shows that if the system's initial position and velocity are described using a joint density and its position coordinates are interpreted angularly, its final position converges, in the weak-star topology, to a distribution uniform on $[0,1]^n$, as the friction coefficient μ tends to zero. Since the coordinates of a distribution uniform on $[0,1]^n$ are independent, this means that no matter how highly correlated the system's initial position and velocity are (as long as they are physically independent), the coordinates describing its final position are approximately independent (in the statistical sense) for small values of the friction coefficient μ. This result is reviewed in Sect. 5.2.

This phenomenon is present in many of the applications considered in Chap. 4. It is also closely related to Hopf's Independence Theorem, a result discussed in Sect. 5.5.

1.8 Statistical Regularity of a Dynamical System

Hopf introduced the concept of statistical regularity to make precise the notion of "unpredictability" of a conservative dynamical system. A dynamical system is said to be *statistically regular* if, as time passes, its position converges to a distribution that does not depend on the joint density describing initial conditions.

Hopf was one of the pioneering researchers in ergodic theory. His work in this area is closely related to that on the method of arbitrary functions. He introduced the notion of strong-mixing for measure-preserving transformations and showed that it is equivalent to statistical regularity. This result is studied in Sect. 5.6

The concept of statistical regularity is extended to an arbitrary dynamical system in Chap. 5. This leads to a notion equivalent to that of mixing sequence of random variables introduced by Rényi and Révész (1958).

A dynamical system that is not statistically regular is the simple harmonic oscillator discussed at the beginning of this introduction. A block of mass m is attached to an ideal spring and free to move over a frictionless horizontal table. The spring is stretched a distance from its equilibrium position and released from rest. If the initial deformation is not too large, Hooke's law applies, and the block's displacement at time t, $x(t)$, is equal to

$$x(t) = A\cos(\omega t),$$

where A denotes the spring's initial displacement and ω its angular frequency. It was shown that the block's displacement, $x(t)$, converges to the product of the random variable A and an independent arcsine law, S. As the limiting random variable does involve A, this dynamical system is not statistically regular. Yet dependence on the spring's angular frequency has washed away. Furthermore, the limit does exhibit some kind of statistical regularity because the scale mixture determined by A always involves the same arcsine law. This phenomenon shows up in many applications of the method of arbitrary functions (cf. the coupled harmonic oscillator discussed above). It motivated generalizing the concept of statistical regularity to that of *partial* statistical regularity. This notion provides a unified framework from which the method of arbitrary functions may be viewed. Some basic characterization theorems for statistically regular systems are extended to the case of partial statistical regularity in Sect. 5.7.

1.9 More Applications

Chapters 3, 4 and 5 include many applications of the method of arbitrary functions that have not been discussed above. Among others, coin tossing, integrable systems (such as a heavy symmetric top), Buffon's needle problem, bouncing balls, billiards, random number generators and gas molecules in a room are considered. These applications may be read independently from the sections containing the mathematical results.

2. Preliminaries

The notation used throughout this work is introduced in Sect. 2.1. In the three sections that follow, various types of convergence of random variables are defined and their main properties derived. Finally, some basic concepts of number theory are introduced in Sect. 2.5.

2.1 Basic Notation

1 The set of real n-tuples is denoted by $\mathbb{R}^n$, in particular the set of real numbers is denoted by $\mathbb{R}$.

2 The set of integer valued n-tuples is denoted by $\mathbb{Z}^n$ while $\mathbb{Z}^n_*$ denotes the set that results from excluding the vector $(0,\ldots,0)$ from $\mathbb{Z}^n$. In particular, $\mathbb{Z}$ denotes the set of integers and $\mathbb{Z}_*$ the set of integers different from zero.

3 The set of positive integers, $\{1,2,3,\ldots\}$, is denoted by $\mathbb{N}$.

4 A distribution uniform on the unit hypercube, $[0,1]^n$, is denoted by U_n.

5 Given a real number x, its integer and fractional parts are denoted by $[x]$ and $x(\mathrm{mod}\,1)$, respectively.

6 All vectors are viewed as column vectors.

7 The superscript $'$ as in x' or A' indicates the transpose of the corresponding vector or matrix.

8 Given an n dimensional vector $x = (x_1,\ldots,x_n)'$, $x(\mathrm{mod}\,1)$ denotes the vector $(x_1(\mathrm{mod}\,1),\ldots,x_n(\mathrm{mod}\,1))'$.

9 When a distribution is said to be absolutely continuous it means that it is absolutely continuous with respect to Lebesgue measure.

10 When referring to a random variable X, it is understood that X is a one dimensional, real valued distribution. The characteristic function of X evaluated at the real number λ is defined as

$$\widehat{f}(\lambda) = \mathrm{E}\,e^{i\lambda X},$$

where E denotes the expectation operator, e the basis of natural logarithms and i the imaginary unit.

11   When referring to a random vector $X = (X_1, \ldots, X_n)'$, it is understood that X is an n dimensional, real valued distribution. The characteristic function of X evaluated at the n dimensional vector $\lambda = (\lambda_1, \ldots, \lambda_n)'$ is defined as

$$\widehat{f}(\lambda) = \mathrm{E}e^{i\lambda \cdot X},$$

where $\lambda \cdot X$ denotes the usual inner product in $\mathbb{R}^n$.

2.2 Weak-star Convergence

Definition. The sequence of n dimensional random vectors $X_1, X_2, \ldots$ with distribution functions $F_1(x), F_2(x), \ldots$ converges in the *weak-star topology* to the random vector X (with distribution function $F(x)$) if $F_k(x)$ converges to $F(x)$ at every point of continuity of F, as k tends to infinity.   □

Assume X is an n dimensional random vector. Then (see e.g. Billingsley, 1986, p.355) the characteristic function of X completely determines its distribution function. This work deals mainly with distributions taking values in $[0,1]^n$. Proposition 2.1.a gives conditions under which these distributions are determined by the values their characteristic function takes at all points of the form $2\pi m$, $m \in \mathbb{Z}^n$. The complex numbers $\widehat{f}(2\pi m)$, $m \in \mathbb{Z}^n$, are called the *Fourier coefficients* of X. Proposition 2.1.b provides conditions under which convergence of the Fourier coefficients of a sequence of random vectors $X_1, X_2, \ldots$ supported by $[0,1]^n$ to the Fourier coefficients of a random vector X supported by $[0,1]^n$ implies weak-star convergence of X_k to X.

Proposition 2.1 *Let X_1, X_2, ... denote a sequence of n dimensional random vectors supported by $[0,1]^n$, and assume that X and Y are two n dimensional random vectors taking values in $[0,1]^n$ and having probability zero of belonging to any $(n-1)$ dimensional face of $[0,1]^n$, that is, to any set of the form $A_1 \times \cdots \times A_n$ where one of the A_i's is equal to $\{0\}$ or $\{1\}$ and the remaining A_j's equal to the unit interval, $[0,1]$. Then:*

a)   *X and Y have the same distribution if and only if they have the same Fourier coefficients.*

b)   *The sequence $X_1, X_2, \ldots$ converges in the weak-star topology to X if and only if given any $m \in \mathbb{Z}^n$ the sequence of Fourier coefficients associated to m converges to the corresponding Fourier coefficient of X.*

Proof. Follows from the n dimensional Weierstrass theorem, see Billingsley (1986, p.361) and Billingsley (1968, p.50ff).   □

Definition. Given two n dimensional random vectors X and Y with distribution functions F_X and F_Y, the Kolmogorov distance between X and Y is defined as

$$d_K(X, Y) = \sup_{x \in \mathbb{R}^n} |F_X(x) - F_Y(x)|. \quad □$$

The following proposition provides conditions under which the Kolmogorov distance metrizes weak-star convergence.

Proposition 2.2 *Assume $X, X_1, X_2, \ldots$ are n dimensional random vectors supported by $[0,1]^n$ and denote their distribution functions by $F(x), F_1(x), F_2(x), \ldots$ If $F(x)$ is continuous then $X_1, X_2, \ldots$ converges in the weak-star topology to X if and only if the Kolmogorov distance between X_k and X tends to zero as k tends to infinity.*

Proof. The non trivial part of the proof is to show that weak-star convergence implies convergence in the Kolmogorov distance. Assume therefore that $X_1, X_2, \ldots$ converges in the weak-star topology to X. As the X_i's and X are supported by $[0,1]^n$, it suffices to show that

$$\lim_{k \to \infty} \sup_{x \in [0,1]^n} |F_n(x) - F(x)| = 0,$$

or equivalently, that

$$(\forall \varepsilon > 0)(\exists k_0 \in \mathbb{N})(\forall k \geq k_0) \sup_{x \in [0,1]^n} |F_k(x) - F(x)| \leq \varepsilon. \tag{2.1}$$

Fix $\varepsilon > 0$. As $F(x)$ is continuous, it is uniformly continuous on $[0,1]^n$ and therefore there exists $\delta > 0$ such that

$$\|x - y\|_\infty \leq \delta \quad \text{implies} \quad |F(x) - F(y)| \leq \frac{\varepsilon}{2}, \tag{2.2}$$

where $\|z\|_\infty$ denotes the largest among the absolute values of the coordinates of the vector z.

Let K be an integer larger than $1/\delta$ and denote by $x_1, \ldots, x_T$ all points of the form $\frac{1}{K}(p_1, \ldots, p_n)'$ with the p_i's in $\{0, 1, \ldots, K\}$. Equation (2.2) implies that

$$\text{If} \quad \|x_i - x_j\|_\infty \leq \frac{1}{K} \quad \text{then} \quad |F(x_i) - F(x_j)| \leq \frac{\varepsilon}{2}. \tag{2.3}$$

As $X_1, X_2, \ldots$ converge in the weak-star topology to X and there is a finite number of x_i's, there exists an integer k_0 such that

$$(\forall k \geq k_0)(\forall i)|F_k(x_i) - F(x_i)| \leq \frac{\varepsilon}{2}. \tag{2.4}$$

Assume $x \in [0,1]^n$. There exist x_i and x_j such that $\|x_i - x_j\|_\infty = 1/K$ and every coordinate of x is larger or equal than the corresponding coordinate of x_i and smaller or equal than the corresponding coordinate of x_j. Therefore for $k \geq k_0$:

$$\begin{aligned}
|F_k(x) - F(x)| &= \max \left(F_k(x) - F(x), \; F(x) - F_k(x) \right) \\
&\leq \max \left(F_k(x_j) - F(x_i), \; F(x_j) - F_k(x_i) \right) \\
&= F(x_j) - F(x_i) + \max \left(F_k(x_j) - F(x_j), \; F(x_i) - F_k(x_i) \right) \\
&\leq \varepsilon,
\end{aligned}$$

where the first inequality follows from the fact that the coordinates of x are larger or equal than those of x_i and smaller or equal than those of x_j and the second one from (2.3) and (2.4). Thus (2.1) follows and the proof concludes.   □

The following proposition provides conditions under which weak-star convergence of a sequence of random vectors conditioned on some random vector imply convergence in the unconditional case.

Proposition 2.3 *Assume* $X, X_1, X_2, \ldots$ *are* n *dimensional random vectors and* Y *is a* p *dimensional random vector such that* $(X|Y), (X_1|Y), (X_2|Y), \ldots$ *have absolutely continuous distributions and* $(X_k|Y = y)$ *converges in the weak-star topology to* X *for* y *in a set to which* Y *assigns probability one. Then, as* k *tends to infinity,* X_k *converges to* X *in the weak-star topology.*

Proof. Follows from the definition of weak-star convergence and Lebesgue's Dominated Convergence Theorem.   □

The following proposition shows how upper bounds for the Kolmogorov distance between $(X_k|Y = y)$ and X may be used to obtain upper bounds for the Kolmogorov distance between X_k and X.

Proposition 2.4 *Assume* X *and* Y *are* n *and* p *dimensional random vectors such that* (X, Y) *is absolutely continuous. Let* Z *denote an absolutely continuous* n *dimensional random vector such that*

$$d_K\left((X|Y = y),\, Z\right) \leq c(y). \tag{2.5}$$

Then

$$d_K(X, Z) \leq \mathrm{E}\big(c(Y)\big).$$

Proof. Denote densities and distribution functions by $f(x)$ and $F(x)$, respectively, using subscripts to indicate the corresponding random vector.

Consider $x = (x_1, \ldots, x_n)' \in \mathbb{R}^n$ and define $A_x = \{y = (y_1, \ldots, y_n)' : y_i \leq x_i; \quad i = 1, \ldots, n\}$. Equation (2.5) implies that

$$\left| \int_{A_x} \big(f_{X,Y}(u, y) - f_Z(u) f_Y(y)\big) du \right| \leq c(y) f(y),$$

and therefore

$$\begin{aligned}
|F_X(x) - F_Z(x)| &= \left| \int_{\mathbb{R}^p} \int_{A_x} \big(f_{X,Y}(u, y) - f_Z(u) f_Y(y)\big) du\, dy \right| \\
&\leq \int_{\mathbb{R}^p} c(y) f_Y(y) dy \\
&= \mathrm{E}\big(c(Y)\big).
\end{aligned}$$

The proof now follows from the fact that $x \in \mathbb{R}^n$ is arbitrary.   □

2.3 Variation Distance

Definition. The *variation distance* between two n dimensional random vectors X and Y, $d_V(X,Y)$, is defined as $\sup_A |\Pr\{X \in A\} - \Pr\{Y \in A\}|$ where the supremum is taken over all Borel sets A in $\mathbb{R}^n$. The sequence X_k converges in the *variation distance* to X if $d_V(X_k,X)$ tends to zero as k tends to infinity.   □

Remark. The Kolmogorov distance between two random vectors is less than or equal to the corresponding variation distance and therefore convergence in the variation distance is a stronger notion than weak-star convergence.   □

In the following proposition, alternative expressions for the variation distance between two absolutely continuous random vectors are stated. The proofs, which are elementary (see Pitman, 1979, p.6), are omitted.

Proposition 2.5 *Assume X and Y are absolutely continuous n dimensional random vectors with corresponding densities $f(x)$ and $g(x)$. Denote the set $\{x : f(x) > g(x)\}$ by A and its complement by A^c. Then*

a) $d_V(X,Y) = \Pr\{X \in A\} - \Pr\{Y \in A\} = \Pr\{Y \in A^c\} - \Pr\{X \in A^c\}$

b) $d_V(X,Y) = \dfrac{1}{2} \displaystyle\int |f(x) - g(x)| dx.$ □

The following proposition shows that, given any measurable function $h(x)$, the variation distance between $h(X)$ and $h(Y)$ cannot exceed that between X and Y.

Proposition 2.6 *If X and Y are n dimensional random vectors and h is a measurable function taking values in any space, then*

$$d_V\big(h(X), h(Y)\big) \leq d_V(X,Y).$$

Proof. Follows directly from the definition of variation distance.   □

A lower bound for the variation distance between two random vectors in terms of their Fourier coefficients is obtained in the following proposition.

Proposition 2.7 *Assume X and Y are n dimensional, absolutely continuous random vectors with densities $f(x)$ and $g(x)$ and corresponding characteristic functions $\widehat{f}(\lambda)$ and $\widehat{g}(\lambda)$. Then:*

$$d_V(X,Y) \geq \frac{1}{2} \sup_\lambda |\widehat{f}(\lambda) - \widehat{g}(\lambda)|.$$

Proof. Due to the definition of a characteristic function and Proposition 2.5.b) it follows that:

$$|\widehat{f}(\lambda) - \widehat{g}(\lambda)| = \left| \int e^{i\lambda \cdot x}(f(x) - g(x))dx \right|$$

$$\leq \int |f(x) - g(x)|dx$$

$$= 2d_V(X,Y). \square$$

The following proposition provides two ways in which upper bounds on the variation distance between conditional distributions lead to bounds in the unconditional case.

Proposition 2.8 *Assume X and Z are absolutely continuous n dimensional random vectors with densities $f_X(x)$ and $f_Z(z)$. Let Y denote a p dimensional random vector with distribution function $F_Y(y)$, and assume X has a density when conditioned on $Y = y$, $f_{X|Y}(x|y)$. Then*

a) $$d_V(X,Z) \leq \int d_V\left((X|Y = y), Z\right)dF_Y(y). \tag{2.6}$$

If Z also has a conditional density with respect to Y, $f_{Z|Y}(z|y)$, then:

b) $$d_V(X,Z) \leq \int d_V\left((X|Y = y), (Z|Y = y)\right)dF_Y(y). \tag{2.7}$$

Proof.

$$d_V(X,Z) = \frac{1}{2} \int |f_X(x) - f_Z(x)|dx$$

$$= \frac{1}{2} \int \left| \int f_{X|Y}(x|y)dF_Y(y) - f_Z(x) \right| dx$$

$$= \frac{1}{2} \int \left| \int \{f_{X|Y}(x|y) - f_Z(x)\}dF_Y(y) \right| dx$$

$$\leq \frac{1}{2} \int \int |f_{X|Y}(x|y) - f_Z(x)|dF_Y(y)\, dx.$$

Applying Fubini's Theorem then leads to

$$d_V(X,Z) \leq \frac{1}{2} \int \left\{ \int |f_{X|Y}(x|y) - f_Z(x)|dx \right\} dF_Y(y)$$

$$= \int d_V\left((X|Y = y), Z\right)dF_Y(y).$$

The proof of (2.7) is similar to that of (2.6); all that needs to be done is replace $f_Z(x)$ by $\int f_{Z|Y}(x|y)dF_Y(y)$ at the appropriate step. $\square$

Remark. Equality holds in (2.6) if Y is sufficient both for X and Z. This result is known as the "sufficiency lemma."

Proposition 2.9 *Assume X_1 and Y_1 are p dimensional random vectors and X_2 and Y_2 are q dimensional random vectors such that (X_1, X_2, Y_1, Y_2) is absolutely continuous. Then:*

$$d_V\left((X_1, X_2), (Y_1, Y_2)\right) \leq E_{X_1} d_V\left((X_2|X_1), (Y_2|X_1)\right) + E_{Y_2} d_V\left((X_1|Y_2), (Y_1|Y_2)\right).$$

Proof.

$$
\begin{aligned}
d_V\left((X_1, X_2), (Y_1, Y_2)\right) &= \frac{1}{2}\int |f_{X_1,X_2}(x_1, x_2) - f_{Y_1,Y_2}(x_1, x_2)|\, dx_1\, dx_2 \\
&\leq \frac{1}{2}\int |f_{X_1,X_2}(x_1, x_2) - f_{X_1,Y_2}(x_1, x_2)|\, dx_1\, dx_2 \\
&\quad + \frac{1}{2}\int |f_{X_1,Y_2}(x_1, x_2) - f_{Y_1,Y_2}(x_1, x_2)|\, dx_1\, dx_2 \\
&= \int \left\{\frac{1}{2}\int |f_{X_2|X_1}(x_2|x_1) - f_{Y_2|X_1}(x_2|x_1)|\, dx_2\right\} f_{X_1}(x_1) dx_1 \\
&\quad + \int \left\{\frac{1}{2}\int |f_{X_1|Y_2}(x_1|x_2) - f_{Y_1|Y_2}(x_1|x_2)|\, dx_1\right\} f_{Y_2}(x_2) dx_2 \\
&= E_{X_1} d_V\left((X_2|X_1), (Y_2|X_1)\right) + E_{Y_2} d_V\left((X_1|Y_2), (Y_1|Y_2)\right).
\end{aligned}
$$

The first equality follows from Proposition 2.5.b) while the remaining steps are straightforward. □

Corollary. *If, in addition to the assumptions of Proposition 2.9, X_1 is independent of (X_2, Y_2) and Y_2 is independent of (X_1, Y_1) then*

$$d_V\left((X_1, X_2), (Y_1, Y_2)\right) \leq d_V(X_1, Y_1) + d_V(X_2, Y_2).$$

Proof. Trivial. □

2.4 Sup Distance

Assume X and Y are absolutely continuous, n dimensional random vectors. Even though the variation distance is a rather strong metric, when interested in the difference between the probability that X belongs to a certain set and the probability that Y belongs to the same set, it does not take into account the set's size. This can be a drawback when small sets are considered. A notion of distance that decreases with the size of the sets under consideration is the sup norm.

Definition. The *sup distance* between two absolutely continuous random vectors X and Y with densities $f(x)$ and $g(x)$, $\|X - Y\|_\infty$, is equal to $\sup_x |f(x) - g(x)|$. The sequence X_k converges to X in the *sup norm* as k tends to infinity if $\|X_k - X\|_\infty$ tends to zero. Strictly speaking, as the density of a random variable is not uniquely defined, the definition should be given in terms of the essential supremum.   □

The following proposition shows how the sup distance takes account of the size of the set A when calculating an upper bound for $|\Pr\{X \in A\} - \Pr\{Y \in A\}|$. Its proof is straightforward and is omitted.

Proposition 2.10 *Assume X and Y are absolutely continuous n dimensional random vectors and A is a bounded measurable subset of $\mathbb{R}^n$ with Lebesgue measure $m(A)$. Then*

$$|\Pr\{X \in A\} - \Pr\{Y \in A\}| \leq m(A)\|X - Y\|_\infty.$$ □

The following proposition implies that when considering absolutely continuous random variables, convergence in the sup distance is a stronger notion than convergence in the variation distance.

Proposition 2.11 *Assume X and Y are absolutely continuous, n dimensional random vectors supported by $[0,1]^n$. Then*

$$d_V(X,Y) \leq \frac{1}{2}\|X - Y\|_\infty.$$

Proof. Follows from Proposition 2.5.b) and the definition of sup distance.   □

2.5 Some Concepts from Number Theory

Notation.   The distance between a real number u and the nearest integer is denoted by $<u>$. That is,

$$<u> = \min_{p \in \mathbb{Z}} |p - u| = \min\left(u(\bmod 1), 1 - u(\bmod 1)\right).$$

From the theory of continued fractions (see Khinchin, 1964, for a beautiful introduction to this topic) it follows that given any irrational number u there exists a sequence of rational numbers p_n/q_n, $n = 1, 2, \ldots$ converging to u such that

$$\left| u - \frac{p_n}{q_n} \right| \leq \frac{1}{q_n^2}. \tag{2.8}$$

A measure of how well an irrational number may be approximated by rational numbers is given by how much larger than 2 the exponent on the right hand side in (2.8) may be chosen. The following definition makes this idea precise.

Definition. The *type* of an irrational number u is defined as

$$\eta(u) = \sup\{\gamma : \liminf_{n \to \infty} n^\gamma <nu> = 0\}.$$

Remark 1. Equation (2.8) implies that $\liminf_{n\to\infty} n^\gamma <nu> = 0$ for any $\gamma < 1$, and therefore the type of every irrational number is larger or equal than one.

Remark 2. Khinchin's Theorem (see Khinchin, 1964) implies that, except for a set of Lebesgue measure zero, all irrational numbers have type equal to one.

Remark 3. Some authors define the type of u as the largest exponent that can be used on the right hand side of (2.8). Their definition of type differs from that introduced here by one.

Remark 4. A celebrated theorem conjectured by Siegel (1921) and proved by Roth (1955) states that all algebraic irrational numbers have type equal to one.

Remark 5. The first numbers known to be transcendental were Liouville's numbers, for example, $\sum_{n=1}^{\infty} 10^{-n!}$. They all have type equal to infinity. $\qquad\square$

Assume u is irrational and define

$$M(u) = \sum_{l\neq 0}\sum_{k\neq 0} \frac{1}{|kl||k + lu|}.$$

Upper bounds for $M(u)$ are used in Chap. 6. In Proposition 2.13 it is shown that $M(u)$ is finite if u has type less than 2. For quadratic irrationals, explicit bounds for $M(u)$ are computed using Propositions 2.13 and 2.14.

The proof of Proposition 2.13 is based on Proposition 2.12 which is due to Hardy and Littlewood (1922). Their original proof has been refined so as to keep track of the bounds.

Proposition 2.12 (Hardy and Littlewood) *Assume u is an irrational number of type less than 2, that is, there exist h (with $1 < h < 2$) and $B > 0$ such that*

$$n^h <nu> > B; \qquad n = 1, 2, \ldots . \tag{2.9}$$

Define the function

$$s(h) = (1 + \log_2 e)\sum_{n\geq 2} \frac{\log n}{(n - 1)^{3-h}} + (\log_2 e)^2 \sum_{n\geq 2}\frac{(\log n)^2}{(n - 1)^{3-h}}. \tag{2.10}$$

Then

$$\sum_{n=1}^{\infty} \frac{1}{n^2 <nu>} \leq \frac{s(h)}{B}. \tag{2.11}$$

Proof. For the sake of clarity, the proof is broken up into several parts.

1.   Let $u_n = 1/(n^h <nu>)$ and $T_m = \sum_{n=m+1}^{2m} u_n$. Define h_n via

$$n^{h_n} <nu> = B.$$

The idea is to partition the terms in T_m into various classes according to their values of h_n and calculate upper bounds for the sum within each class.

2.   Due to (2.9), the sum of terms in T_m for which $h_n \le h - 1$ is bounded by $\frac{1}{B} \sum_{n=m+1}^{2m} \frac{1}{n}$.

3.   The remaining terms in T_m are classified as follows. Given a positive integer η define

$$b_r = h - 1 + \frac{r}{\eta}; \qquad r = 0, 1, \ldots, \eta - 1,$$

and call a typical term u_n of T_m a term of class r if $b_r \le h_n < b_{r+1}$.

It is now shown that for any u_n of class r, u_{n+s} has to be of class less than r for $1 \le s \le 2n^{b_r/h}$.

First note that, due to (2.9), for such an s:

$$< su >> Bs^{-h} \ge 2Bn^{-b_r}. \tag{2.12}$$

As u_n is of class r, $< nu > \le Bn^{-b_r}$. This combined with (2.12) and the general fact that if $<v>> 2c$ and $<w> \le c$ then $<v+w>> c$, leads to

$$<(n+s)u >> Bn^{-b_r} > B(n+s)^{-b_r}.$$

Therefore u_{n+s} is of class less than r.

4.   Next it is shown that the number of terms of class r is at most $m^{1-\frac{b_r}{h}}$.

From part 3 it follows that for every term in T_m of class r there exist (at least) $2m^{b_r/h}$ terms in T_m that are not of class r, except possibly for the last term of class r that participates in T_m. Hence there are at most $\frac{1}{2}m^{1-\frac{b_r}{h}} + 1$ terms of class r.

Note that

$$m^{1-\frac{b_r}{h}} = m^{\frac{1}{h} - \frac{r}{\eta h}} \ge m^{1/\eta h} > 1,$$

and therefore $m^{1-\frac{b_r}{h}}$ is an upper bound for the number of terms of class r in T_m.

5.   It is now shown that the contribution of terms of class r to T_m is at most $\frac{1}{B}m^{1/\eta}$.

Due to (2.9) and the definition of u_n it follows that if u_n participates in T_m and is of class r then

$$u_n = \frac{1}{B}n^{h_n - h} < \frac{1}{B}n^{b_{r+1} - h} \le \frac{1}{B}m^{b_{r+1} - h},$$

and therefore – due to part 4 – the contribution of all terms of class r in T_m is at most $\frac{1}{B}m^{b_{r+1} - h + 1 - \frac{b_r}{h}}$. But

$$b_{r+1} - h + 1 - \frac{b_r}{h} = \frac{1}{\eta} - \left(1 - \frac{1}{h}\right)\left(1 - \frac{r}{\eta}\right) \le \frac{1}{\eta},$$

where the last step follows from the fact that $h \ge 1$. Hence the contribution of all terms of class r to T_m is bounded by $\frac{1}{B}m^{1/\eta}$.

6.   Parts 2 and 5 with η equal to the integer part of the logarithm in base 2 of m, that is, $\eta = [\log_2 m]$, imply that

$$T_m \leq \frac{1}{B}\left\{\left(\sum_{n=m+1}^{2m}\frac{1}{n}\right) + [\log_2 m]m^{1/[\log_2 m]}\right\}. \qquad (2.13)$$

7.   Let $U_l = \sum_{k=1}^{l} u_k$. Then

$$U_{2^l} = u_1 + \sum_{k=0}^{l-1} T_{2^k}.$$

This combined with (2.13) and the trivial bound for $\sum 1/n$ in terms of $\int x^{-1} dx$ implies that

$$U_{2^l} \leq \frac{1}{<u>} + \frac{l}{B}\{\log 2 + l - 1\}, \qquad (2.14)$$

where log without subindex denotes the natural logarithm. Given an arbitrary positive integer m, let $l = [\log_2 m] + 1$ in (2.14) to conclude that

$$U_m \leq \frac{1}{<u>} + \frac{(\log_2 m + 1)}{B}\{\log 2 + \log_2 m\}. \qquad (2.15)$$

8.   Abel's Summation Formula and (2.15) imply that

$$\sum_{n=1}^{m}\frac{1}{n^2 <nu>} \leq \frac{(\log 2 + 1)}{B}\sum_{n=2}^{m}\left(\frac{1}{(n-1)^{2-h}} - \frac{1}{n^{2-h}}\right)\log_2 n$$

$$+\frac{1}{B}\sum_{n=2}^{m}\left(\frac{1}{(n-1)^{2-h}} - \frac{1}{n^{2-h}}\right)(\log_2 n)^2 + o(1), \qquad (2.16)$$

where $o(1)$ denotes terms that tend to zero as m tends to infinity. As $1 < h < 2$:

$$\frac{1}{(n-1)^{2-h}} - \frac{1}{n^{2-h}} \leq \frac{1}{(n-1)^{3-h}},$$

and therefore (2.16) leads to (2.11). The proof concludes.   □

Lemma 2.1 *Given any real number $v > 1$ define*

$$S(v) = \sum\nolimits^{*} \frac{1}{|k(k+v)|},$$

where $\sum^{*}$ indicates that the sum is taken over all integers k different from zero such that $|k+v| > 1$. Then

$$S(v) \leq \frac{2}{v} + \frac{4\log v}{v} + \frac{2}{[v]-1}I\{v > 2\},$$

where $[v]$ denotes the fractional part of v and I an indicator function.

Proof. The sum defining $S(v)$ is broken up into four pieces, in such a way that the function $x \to 1/|x(x+v)|$ is monotone in every region. This results in summing over

$k \geq 1$, $[-v/2] + 1 \leq k \leq -1$, $[-v] + 1 \leq k \leq [-v/2]$ and $k \leq -[v] - 2$. Every sum is now bounded comparing it with $\int dx/|x(x+v)|$. For example:

$$\sum_{k=-\infty}^{-[v]-2} \frac{1}{|k(k+v)|} \leq \int_{-\infty}^{-[v]-2} \frac{dx}{x(x+v)} + \frac{1}{([v]+2)(2-\{v\})}$$

$$\leq \int_{-\infty}^{-v-1} \frac{dx}{x(x+v)} + \frac{1}{2+[v]}$$

$$= \frac{\log(1+v)}{v} + \frac{1}{2+[v]}, \tag{2.17}$$

where $\{v\}$ denotes the integer part of v.

Next note that the sums for $-[v] + 1 \leq k \leq -1$ only make sense if $v > 2$. In this case, for example,

$$\sum_{k=-[v]+1}^{[-v/2]} \frac{1}{k(k+v)|} \leq \frac{1}{([v]-1)(\{v\}+1)} + \int_{1-[v]}^{[-v/2]} \frac{dx}{x(-x-v)}$$

$$\leq \frac{1}{[v]-1} + \int_{1-v}^{-v/2} \frac{dx}{x(-x-v)}$$

$$= \frac{1}{[v]-1} + \frac{\log(v-1)}{v}. \tag{2.18}$$

Similar calculations lead to

$$\sum_{k=[-v/2]+1}^{-1} \frac{1}{|k(k+v)|} \leq \frac{\log(v-1)}{v} + \frac{1}{v-1}, \tag{2.19}$$

and

$$\sum_{k\geq 1} \frac{1}{|k(k+v)|} \leq \frac{\log(1+v)}{v} + \frac{1}{1+v}. \tag{2.20}$$

Equations (2.17), (2.18), (2.19) and (2.20) and the fact that $1/(2+[v]) \leq 1/(1+v)$ and $1/(v-1) \leq 1/([v]-1)$ lead to

$$S(v) \leq \frac{2}{1+v} + \frac{2\log(1+v)}{v} + 2\left(\frac{1}{[v]-1} + \frac{\log(v-1)}{v}\right) I\{v > 2\}$$

$$\leq \frac{2}{v} + \frac{2\log(v+1)}{v} + \frac{2\log(v-1)}{v} + \frac{2}{[v]-1} I\{v > 2\}$$

$$\leq \frac{2}{v} + \frac{4\log v}{v} + \frac{2}{[v]-1} I\{v > 2\}$$

which concludes the proof.   □

Proposition 2.13 *Let u be an irrational number of type less than 2. Therefore there exists h in $(1,2)$ and $B > 0$ such that*

$$n^h <nu>> B; \qquad n = 1, 2, \dots.$$

Define

$$M(u) = \sum_{l \neq 0} \sum_{k \neq 0} \frac{1}{|kl||k + lu|}.$$

Assume $u > 1$. Note that $M(u) = M(-u) = M(1/u)/u$ and hence upper bounds for $M(u)$ when $u > 1$ lead to upper bounds for any u. Then:

$$M(u) \leq \frac{2s(h)}{(u-1)B} + \frac{2\pi^2}{3} \left(\frac{1}{u} + \frac{2}{u-1} + \frac{2\log u}{u} \right) +$$

$$\frac{4}{u} \left(\frac{3}{2} \log 2 + 1 \right) + \frac{4}{[u]-1} I\{u > 2\} - \frac{4}{u-1},$$

where $[u]$ denotes the integer part of u, I an indicator function and $s(h)$ the function defined in Proposition 2.12 (see (2.10)).

Proof. First note that, as the terms associated to (k, l) and $(-k, -l)$ in the sum that defines $M(u)$ are the same,

$$M(u) = 2 \sum_{l=1}^{+\infty} \sum_{k \neq 0} \frac{1}{|kl||k + lu|}. \tag{2.21}$$

Next denote by k_0 the integer that minimizes $|k + lu|$ over all integers k, for fixed l, $l \geq 1$. As $lu > 1$, $|k_0|$ is equal to either $[lu]$ or $[lu] + 1$. Hence

$$k_0 \geq [lu] \geq lu - 1 \geq l(u - 1) > 0,$$

where the last two inequalities used the facts that $l \geq 1$ and $u > 1$. Therefore

$$\sum_{k \neq 0} \frac{1}{|k(k + lu)|} \leq \frac{1}{l(u-1) <lu>} + \frac{2}{lu-1} + S(lu), \tag{2.22}$$

with the first term corresponding to k_0, the second one to the second nearest integer to lu (it has to be at least one half apart) and $S(lu)$ defined in Lemma 2.1. Also note that if $l \geq 2$ then

$$[lu] - 1 \geq lu - 2 \geq lu - l = l(u - 1) \geq 0,$$

and therefore

$$\sum_{l \geq 1} \frac{I\{lu > 2\}}{l([lu] - 1)} \leq \frac{I\{u > 2\}}{[u] - 1} + \frac{1}{u-1} \sum_{l \geq 2} \frac{1}{l^2}. \tag{2.23}$$

Substituting (2.22) into (2.21), applying Lemma 2.1, (2.23) and the fact that $\sum_{l \geq 1} 1/l^2 = \pi^2/6$ and

$$\sum_{l \geq 2} \frac{\log l}{l^2} \leq \frac{\log 2}{4} + \int_2^{+\infty} \frac{\log x}{x^2} dx$$

$$= \frac{3}{4} \log 2 + \frac{1}{2},$$

leads to the required bound.   □

Proposition 2.14 (Liouville) *Assume u is a quadratic irrational, i.e. it is irrational and there exist integers a_0, a_1, a_2; $a_2 \neq 0$, such that u is a root of*

$$a_2 x^2 + a_1 x + a_0 = 0. \tag{2.24}$$

Let v denote the other root of (2.24) and define c by

$$c^{-1} = |a_2|(|v - u| + 1). \tag{2.25}$$

Given h in the interval $(1, 2)$ let

$$n_0 = 1/c^{1/(h-1)},$$

and define

$$B = \min\{n^h <nu>; \ 1 \leq n \leq n_0\}.$$

Then

$$n^h <nu> \geq B ; \qquad n = 1, 2, \ldots \tag{2.26}$$

Proof. The proof follows from the classical proof of Liouville's Theorem.

Let $f(x) = a_2 x^2 + a_1 x + a_0$. As u (and therefore v) are irrational, $n^2 f(\frac{m}{n})$ is an integer different from zero for any rational number m/n and therefore

$$1 \leq \left| n^2 f\left(\frac{m}{n}\right) \right| = n^2 |a_2| \left| u - \frac{m}{n} \right| \left| v - \frac{m}{n} \right|. \tag{2.27}$$

If $|u - \frac{m}{n}| \leq 1$, (2.27) implies that $|u - \frac{m}{n}| \geq c/n^2$ (with c defined in (2.25)) and therefore, as $c \leq 1$, for any rational $\frac{m}{n}$:

$$\left| u - \frac{m}{n} \right| \geq \frac{c}{n^2}. \tag{2.28}$$

But $n^h <nu> < 1$ is equivalent to having $|u - \frac{m}{n}| < 1/n^{1+h}$ and due to (2.28) this can only happen if $n^{h-1} < c^{-1}$. This implies that $n^h <nu> \geq 1$ for $n \geq n_0$ and (2.26) follows.   □

Remark 1. Propositions 2.13 and 2.14 imply that upper bounds for $M(u)$ may be effectively computed for any quadratic irrational u.

Remark 2. Note that the constant B in (2.26) depends on h and u. For fixed u it grows with h. It can now be seen that the effect of different values of h for the upper bound derived in Proposition 2.13 is not obvious, because both the numerator and the denominator of $s(h)/B$ grow as h does.

Example. The quadratic irrational $(1 + \sqrt{5})/2$ is a root of $x^2 - x - 1 = 0$, the other root being $(1 - \sqrt{5})/2$. To calculate B, a value of h in $(1, 2)$ has to be chosen. From a practical point of view there are numerical problems if n_0 is too large and therefore h is not chosen near one. For example, if $h = 1.1$, n_0 is larger than 100,000 and the accuracy in evaluating $n^h <nu>$ for such large n is not reliable. A reasonable value of h is 1.3. Then $n_0 = 355$ and $B = 0.618$. The value of $s(h)$ at $h = 1.3$ was calculated numerically leading to 16.1. Proposition 2.13 now implies that $M(u) \leq 165$ when $u = (1 + \sqrt{5})/2$.

3. One Dimensional Case

This chapter considers physical systems with one degree of freedom. The fractional part of the product of a large real number t and a random variable X, $(tX)(\bmod 1)$, is studied in detail in Sect. 3.1. The random variable $(tX)(\bmod 1)$ converges in the weak-star topology to a distribution uniform on $[0,1]$ if and only if the characteristic function of X vanishes at infinity (Theorem 3.2, Kemperman). A necessary and sufficient condition for convergence in the variation distance is that X have a density (Theorem 5.3). Convergence may take place at very slow rates (Proposition 3.8). Yet if the density of X has bounded variation, convergence is at a rate at least linear in t^{-1} (Theorem 3.9, Kemperman). Convergence may take place at faster rates for specific families of random variables (Proposition 3.11). Furthermore, there exists a family of random variables for which $(tX)(\bmod 1)$ is identically uniform once t passes a certain threshold (Theorem 3.14).

The results of Sect. 3.1 are applied to various examples in Sect. 3.2. They include bouncing balls, coin tossing, throwing a dart at a wall, Poincaré's roulette problem, Poincaré's Law of Small Planets and an example from the dynamical systems literature. The mathematics developed in Sect. 3.1 is also applied when studying Hopf's contribution to the method of arbitrary functions in Chap. 5.

The section with applications may be read independently from the one containing the mathematical results.

3.1 Mathematical Results

3.1.1 Weak-star Convergence

In Poincaré's analysis of roulette (see Sect. 3.2.4) the mathematical problem considered is finding conditions on the random variable X under which

$$\lim_{n\to\infty} \Pr\left\{(nX)(\bmod 1) \leq \frac{1}{2}\right\} = \frac{1}{2}, \tag{3.1}$$

where n denotes a positive integer.

Poincaré (1896) showed that (3.1) holds if X has a density with bounded derivative. Borel (1909) extended this result to the case of continuous densities. Fréchet (1921) proved (3.1) for any random variable X with a density.

Assume X and Y are real valued random variables with continuous joint density. While deriving his Law of Small Planets (see Sect. 3.2.5), Poincaré (1896) showed that $(tX + Y)(\mathrm{mod}\,1)$ converges, in the weak-star topology, to a distribution uniform on the unit interval as t tends to infinity.

Necessary and sufficient conditions for weak-star convergence of $(tX)(\mathrm{mod}\,1)$ to a uniform distribution are derived in Theorem 3.2. The result is due to Kemperman (1975).

Lemma 3.1 *Assume X is a random variable with distribution function F and characteristic function $\widehat{f}$. Let $\widehat{f}_t$ denote the characteristic function of $(tX)(\mathrm{mod}\,1)$. Then:*

$$\widehat{f}_t(2\pi m) \;=\; \widehat{f}(2\pi mt), \qquad\qquad m = 0, \pm 1, \pm 2, \ldots$$

Proof. It suffices to consider the case $t = 1$:

$$
\begin{aligned}
\widehat{f}_1(2\pi m) &= \mathrm{E}\left(e^{i2\pi m X(\mathrm{mod}\,1)}\right)\\[2mm]
&= \sum_k \int_{[k,k+1)} e^{i2\pi m x(\mathrm{mod}\,1)}\,dF(x)\\[2mm]
&= \sum_k \int_{[k,k+1)} e^{i2\pi m(x-k)}\,dF(x)\\[2mm]
&= \sum_k \int_{[k,k+1)} e^{i2\pi m x}\,dF(x)\\[2mm]
&= \widehat{f}(2\pi m). \qquad\Box
\end{aligned}
$$

Theorem 3.2 (Poincaré, Borel, Fréchet, Kemperman) *Let X be a real valued random variable with characteristic function $\widehat{f}(t)$ and denote by U a distribution uniform on the unit interval. Then $(tX)(\mathrm{mod}\,1)$ converges to U in the weak-star topology, when t tends to infinity, if and only if $\lim_{|t|\to\infty} \widehat{f}(t) = 0$.*

Proof. Proposition 2.1 and the fact that the non trivial Fourier coefficients of U are zero imply that a necessary and sufficient for convergence of $(tX)(\mathrm{mod}\,1)$ to U, in the weak-star topology, is that the corresponding Fourier coefficients converge:

$$\lim_{t\to\infty} \widehat{f}(2\pi mt) = 0 \qquad\qquad m = 1, 2, \ldots$$

Since t can take any real value, this is equivalent to having $\widehat{f}$ vanish at infinity. $\Box$

Remark. From the proof it can be seen that a necessary and sufficient condition for convergence of $(tX)(\mathrm{mod}\,1)$ to U is that one Fourier coefficient of $(tX)(\mathrm{mod}\,1)$ tend to zero as t tends to infinity.

Corollary 1 *If X is a random variable which has a density (with respect to Lebesgue measure) then $(tX)(\bmod 1)$ converges in the weak-star topology to a distribution uniform on the unit interval.*

Proof. Follows directly from the Riemann-Lebesgue Lemma (see Billingsley, 1986, p.354) and Theorem 3.2.   □

Remark. Let b be an algebraic number of order n, that is, b is a root of an equation of the form

$$a_n x^n + a_{n-1} x^{n-1} + \ldots + a_1 x + a_0 = 0,$$

with $a_0, \ldots, a_n$ integers and $a_n \neq 0$. Furthermore, there does not exist any such equation of order less than n that has b as a root. The number b is said to be Pisot-Vijayaraghavan number if $|b| > 1$ and the remaining roots of the preceding equation lie inside the unit circle. The rational numbers satisfy this definition by default. Since the algebraic numbers are countable, the set of Pisot-Vijayaraghavan numbers is countable and has zero Lebesgue measure.

Pisot-Vijayaraghavan numbers may be used to construct an example where the random variable $(tX)(\bmod 1)$ converges weak-star to a distribution uniform on $[0,1]$ even though X is singular continuous (and therefore does not have a density). Let $X = \sum_{i \geq 1} a^i \varepsilon_i$, where $0 < a < 1$, and the ε_i's are independent, identically distributed taking value 0 or 1 with probability $\frac{1}{2}$ (so X is the stationary distribution of the Markov chain $X_0 = 0, X_n = aX_{n-1} + \varepsilon_n$). The characteristic function of X vanishes at infinity unless a is the reciprocal of a Pisot-Vijayaraghavan number. Yet X is singular continuous for $0 < a < \frac{1}{2}$. For a review article on this topic see Garsia (1962).

Corollary 2 *Assume X and Y are random variables such that the conditional distribution of X given $Y = y$ is well defined and has a characteristic function that vanishes at infinity for y in a set to which Y assigns probability one. Then $(tX + Y)(\bmod 1)$ converges, in the weak-star topology, to a distribution uniform on the unit interval, as t tends to infinity.*

Proof. Follows from Theorem 3.2, Proposition 2.3 and Lebesgue's Dominated Convergence Theorem.   □

Remark. The Corollary's hypotheses hold when (X,Y) is absolutely continuous. They also hold if the characteristic function of X vanishes at infinity and Y is an arbitrary random variable, independent of X.   □

There are many ways of making mathematically precise the idea that the distribution of $(tX)(\bmod 1)$ does not depend much on that of X when t is large. Theorem 3.2 did this by showing that for any random variable with a characteristic function vanishing at infinity the distribution of $(tX)(\bmod 1)$ is approximately uniform. An alternative approach consists in showing that $(tX)(\bmod 1)$ and X become approximately independent as t grows. This fact is rather surprising because $(tX)(\bmod 1)$ is a function of X and hence highly dependent on it. These ideas are stated precisely in the following theorem.

Theorem 3.3 *Assume X and Y are random variables and Z is a random vector such that (X,Y,Z) is absolutely continuous. Then, as t tends to infinity, $(tX+Y)(\mathrm{mod}\,1)$ converges in the weak-star topology to a distribution uniform on $[0,1]$ that is independent of (X,Y,Z).*

Proof. This theorem is a particular case of Theorem 4.4 which is proved in the next chapter. □

3.1.2 Bounds on the Rate of Convergence

Convergence of $(tX)(\mathrm{mod}\,1)$ to a distribution uniform on the unit interval, U, can be very slow. If X has a Gamma density which behaves like x^{a-1} near the origin, $a > 0$, the variation distance between $(tX)(\mathrm{mod}\,1)$ and U, $d_V((tX)(\mathrm{mod}\,1),\,U)$, tends to zero at rate t^{-a} (see Sect. 2.3 for the definition and main properties of the variation distance).

In Proposition 3.4 it is shown that $(tX)(\mathrm{mod}\,1)$ has a density if X has one. This result is used in Proposition 3.6 to prove that the rate at which $(tX)(\mathrm{mod}\,1)$ converges to U is very slow for positive random variables with monotone densities which blow up sufficiently fast at the origin. The case of a Gamma random variables with densities blowing up at the origin is considered in Proposition 3.7.

Proposition 3.4 *Let X be a random variable with density $f(x)$. Then*

$$f_t(x) = \frac{1}{t}\sum_{k=-\infty}^{+\infty} f\left(\frac{x+k}{t}\right), \qquad 0 \le x \le 1, \tag{3.2}$$

defines a density for $(tX)(\mathrm{mod}\,1)$.

Proof. Without loss of generality, it suffices to consider the case $t=1$.

If $0 \le c < d \le 1$, Tonelli's Theorem (see e.g. Billingsley, 1986, p.238) implies that:

$$\int_c^d f_1(x)dx = \sum_k \int_c^d f(x+k)dx$$
$$= \sum_k \Pr\{k+c \le X \le k+d\}$$
$$= \Pr\{c \le X(\mathrm{mod}\,1) \le d\},$$

and therefore $f_1(x)$ is a density for $X(\mathrm{mod}\,1)$. □

Lemma 3.5 *Assume X is a positive random variable with monotone density $f(x)$ and distribution function $F(x)$. Then:*

$$0 \le \sup_{0\le x\le1}\left\{F\left(\frac{x}{t}\right) - xF\left(\frac{1}{t}\right)\right\} \le d_V((tX)(\mathrm{mod}\,1),\,U) \le F\left(\frac{1.25}{t}\right).$$

Proof. Let $F_t(x)$ denote the distribution function of $(tX)(\text{mod }1)$. Then

$$F_t(x) - x = \sum_{k \geq 0} \Pr\{k \leq tX \leq k + x\} - x$$

$$= F\left(\frac{x}{t}\right) + \sum_{k \geq 1} \int_{k/t}^{(k+x)/t} f(u)du - x$$

$$\geq F\left(\frac{x}{t}\right) + x \int_{1/t}^{+\infty} f(u)du - x$$

$$= F\left(\frac{x}{t}\right) - xF\left(\frac{1}{t}\right)$$

$$\geq 0.$$

The first inequality follows from the fact that $\int_{k/t}^{(k+x)/t} f(u)du$ is larger or equal than $x \int_{k/t}^{(k+1)/t} f(u)du$ because $f(u)$ is decreasing. The second inequality is obtained by applying the concavity of $F(x)$ and the fact that $F(0) = 0$.

A lower bound for $d_V\left((tX)(\text{mod }1), U\right)$ is provided by the latter inequality. A similar argument, based again on the monotonicity of $f(x)$, shows that

$$1 - F\left(\frac{x+1}{t}\right) \leq \frac{1}{t}\sum_{k \geq 1} f\left(\frac{x+k}{t}\right) \leq 1 - F\left(\frac{x}{t}\right),$$

and combining this with Proposition 3.4 leads to:

$$-F\left(\frac{x+1}{t}\right) + \frac{1}{t}f\left(\frac{x}{t}\right) \leq f_t(x) - 1 \leq -F\left(\frac{x}{t}\right) + \frac{1}{t}f\left(\frac{x}{t}\right).$$

Therefore

$$d_V\left((tX)(\text{mod }1), U\right) = \frac{1}{2}\int_0^1 |f_t(x) - 1|dx$$

$$\leq \frac{1}{2t}\int_0^1 f\left(\frac{x}{t}\right)dx + \frac{1}{2}\int_0^1 F\left(\frac{x+1}{t}\right)dx$$

$$= \frac{1}{2}\int_0^{1/t} f(u)du + \frac{t}{2}\int_{1/t}^{2/t} F(u)du$$

$$\leq \frac{1}{2}F\left(\frac{1}{t}\right) + \frac{1}{2}F\left(\frac{1.5}{t}\right)$$

$$\leq F\left(\frac{1.25}{t}\right),$$

where the concavity of $F(x)$ was used repeatedly.   □

Proposition 3.6 *Assume X is a positive random variable with monotone, continuously differentiable density $f(x)$. Suppose that $f(x)$ is convex in a neighborhood of the origin and that it blows up at the origin sufficiently fast so that*

$$\lim_{u \to 0+} u f'(u) = -\infty. \tag{3.3}$$

Then $d_V\big((tX)(\operatorname{mod}1),\, U\big)$ converges to zero at a rate slower than linear in t^{-1}.

Proof. From Lemma 3.5 it suffices to show that

$$\lim_{t \to \infty} t \left\{ F\left(\frac{0.5}{t}\right) - 0.5 F\left(\frac{1}{t}\right) \right\} = +\infty,$$

or equivalently

$$\lim_{u \to 0+} \frac{F(0.5u) - 0.5F(u)}{u} = +\infty.$$

Monotonicity, convexity and differentiability of $f(x)$ imply that

$$\frac{F(0.5u) - 0.5F(u)}{u} \geq \frac{1}{2u} \int_0^{0.5} \big(f(v) - f(0.5u)\big)\, dv$$

$$\geq -\frac{1}{16} u f'(0.5u).$$

The result now follows from equation (3.3). □

When X has a Gamma density that blows up at the origin, Proposition 3.6 implies that $(tX)(\operatorname{mod}1)$ converges to a uniform distribution at a rate slower than t^{-1}. Lemma 3.5 is now used to establish the exact rate of convergence.

Proposition 3.7 *Assume X has a Gamma density:*

$$f(x) = \frac{1}{\Gamma(a)b^a} x^{a-1} e^{-x/b}\,; \qquad x > 0,$$

with $b > 0, 0 < a < 1$.

Then $d_V\big((tX)(\operatorname{mod}1),\, U\big)$ tends to zero at rate t^{-a}.

Proof. From the fact that $1 - x \leq e^{-x} \leq 1$ it follows that for any $c > 0$

$$\frac{c^a t^{-a}}{\Gamma(a)b^a} \left(\frac{1}{a} - \frac{c}{bt}\right) \leq F\left(\frac{c}{t}\right) \leq \frac{c^a t^{-a}}{\Gamma(a+1)b^a}.$$

These inequalities combined with Lemma 3.5 show that the exact rate for a Gamma density is t^{-a}. □

Remark. The case of a Gamma density that is bounded is considered following Proposition 3.12. □

The previous example shows that convergence of $(tX)(\operatorname{mod}1)$ to a uniform random variable may take place at rates much slower than linear in t^{-1}.

In Theorem 3.9, rates of convergence at least linear in t^{-1} are obtained under the assumption that X has a density with bounded variation. This result is due to Kemperman (1975).

The total variation of a random variable with density $f(x)$ plays an important role in the bounds obtained in Theorem 3.9. In general, it is defined as

$$V(X) = \sup_{x_0 < \ldots < x_n} \sum_{k=1}^{n} |f(x_k) - f(x_{k-1})|,$$

where the supremum is taken over all positive integers n and all partitions $x_0 < \ldots < x_n$ of the real line. The density $f(x)$ (or equivalently the random variable X) is said to be of bounded variation if the total variation of X is finite.

In the following proposition some basic properties of a random variable's total variation are derived.

Proposition 3.8: Properties of a Random Variable's Total Variation *Let X denote a random variable with density $f(x)$ of bounded variation.*

a) *For $t \neq 0$: $V(tX) = \frac{1}{|t|}V(X)$ and for all real numbers a: $V(X + a) = V(X)$.*

b) *If $f(x)$ is piecewise continuously differentiable, with jumps of absolute magnitude $\delta_1, \ldots, \delta_n$, then $V(X) = \sum_{i=1}^{n} \delta_i + \int |f'(x)| dx$. In particular, if $f(x)$ is continuous and piecewise differentiable, $V(X) = \int |f'(x)| dx$.*

c) *If $f(x)$ is unimodal and its maximum value is M, then $V(X) = 2M$. More generally, if $f(x)$ has r modes and the corresponding local maximum and minimum values are $M_1, \ldots, M_r$ and $m_1, \ldots, m_{r-1}$ respectively, then $V(X) = 2\sum_{k=1}^{r} M_k - 2\sum_{k=1}^{r-1} m_k$.*

d) *Assume $f(x)$ is continuously differentiable on the real line and supported by $[a, b]$, $-\infty \leq a \leq b \leq +\infty$. Let $g : [a, b] \to [c, d]$ be one to one, onto, with both itself and its inverse twice continuously differentiable, $-\infty \leq c \leq d \leq +\infty$. Suppose that*

$$\lim_{u \to a+} \frac{f(u)}{|g'(u)|} = 0 = \lim_{u \to b-} \frac{f(u)}{|g'(u)|}.$$

Then:

$$V\big(g(X)\big) \leq \int_a^b \left| \frac{f'(x)}{g'(x)} \right| dx + \int_a^b \frac{|g''(x)|}{|g'(x)|^2} f(x) dx.$$

e) *Assume X is a mixture of random variables belonging to a certain family, with mixing distribution Y. That is,*

$$f_X(x) = \int f_{X|Y}(x|y) dF_Y(y), \qquad (3.4)$$

where it has been assumed that $(X|Y = y)$ has a density, but not necessarily Y. Then the total variation of X is less than or equal to the mean-conditional variation of X given Y:

$$V(X) \leq E_Y V(X|Y).$$

In particular, if all conditional densities $(X|Y = y)$ have total variation smaller than M then the total variation of X is also bounded by M.

Proof. The proofs of parts a), b), c) and d) follow from straightforward calculus.

Part e) can be proved as follows: Consider arbitrary numbers $x_0, \ldots, x_n$ with $x_0 < \ldots < x_n$. Equation (3.4) implies that

$$\sum |f(x_k) - f(x_{k-1})| \leq \int \sum |f_{X|Y}(x_k|y) - f_{X|Y}(x_{k-1}|y)| dF_Y(y)$$

$$\leq \int V(X|Y = y) dF_Y(y),$$

where the last inequality follows from the definition of total variation. The proof concludes by taking supremum over all $x_0 < \ldots < x_n$. $\qquad\square$

Example 1. Assume X is a Gamma random variable with density

$$f(x) = \frac{1}{\Gamma(a)b^a} x^{a-1} e^{-x/b} \; ; \; x > 0,$$

where $a > 0$ and $b > 0$.

From the definition of total variation it follows that $V(X) = +\infty$ if $0 < a < 1$. Proposition 3.8.c) can be used to show that

$$V(X) = \frac{2(a-1)^{a-1} e^{-(a-1)}}{b\Gamma(a)}$$

if $a > 1$ and $V(X) = 2/b$ if $a = 1$ (case of an exponential random variable).

Example 2. Proposition 3.8.c) implies that a normal random variable with variance σ^2 (and any mean) has total variation equal to $2/(\sigma\sqrt{2\pi})$. $\qquad\square$

The random variable $(tX)(\mathrm{mod}\, 1)$ converges in the variation distance to a distribution uniform on the unit interval if and only if X has a density. The proof of this result follows from some general theory discussed in Chap. 5 (see Theorem 5.3). Yet Propositions 3.4, 3.6 and 3.7 show that additional smoothness assumptions are needed to have good rates of convergence.

In the following theorem it is shown that $(tX)(\mathrm{mod}\, 1)$ converges to a distribution uniform on the unit interval at a rate at least linear in t^{-1} if X has bounded variation.

For ease of future reference, various similar results are established separately and designated Theorem 3.9.a), 3.9.b) and 3.9.c) respectively. From a mathematical point of view, all the results that follow are corollaries of Theorem 3.9.a).

Theorem 3.9.a) (Kemperman) *Let X denote a random variable whose density $f(x)$ has bounded variation $V(X)$. Denote by F_t the distribution function of $(tX)(\mathrm{mod}\, 1)$ and let A be a subset of $[0,1]$ with Lebesgue measure L. Then:*

$$|F_t(A) - L| \leq \frac{V(X)L(1-L)}{2t}.$$

Proof. Let $h(x)$ denote a bounded periodic function of period one which satisfies $\int_0^1 h(x)dx = 0$.

Define $H(x) = \int_0^x h(u)du$. Since $f(x)$ is integrable, there exist sequences $x_1, x_2, \ldots$; $y_1, y_2, \ldots$ diverging to $-\infty$ and $+\infty$ respectively, such that $f(x_k)$ and $f(y_k)$ tend to zero. Integration by parts may be applied because $f(x)$ has bounded variation (see Hewitt and Stromberg, 1965, Remark 21.68, p.420) and therefore:

$$Eh(X) = \int h(x)f(x)dx$$
$$= -\lim_{k\to\infty} \int_{x_k}^{y_k} H(x)df(x)$$
$$= -\lim_{k\to\infty} \int_{x_k}^{y_k} (H(x) - d)df(x),$$

where d can be any real number. Choosing d as the midrange of H, that is the average of its maximum and minimum values, gives

$$|Eh(X)| \le \frac{V(X)\mathrm{Osc}(H)}{2},$$

where $\mathrm{Osc}(H)$ denotes the difference between the maximum and minimum values taken by H. Hence

$$|Eh(tX)| \le \frac{V(X)\mathrm{Osc}(H)}{2t}. \tag{3.5}$$

Let $I_1, I_2, \ldots$, be a countable collection of intervals in $[0,1]$ whose total length is L and define

$$h(x) = \begin{cases} 1 - L, & x(\mathrm{mod}\,1) \in \cup I_i; \\ -L, & \text{otherwise}. \end{cases}$$

Then $\mathrm{Osc}(H) \le L(1 - L)$ since no matter where H attains its maximum, it cannot decrease more than L times the length of the complement of $\cup I_i$ with respect to $[0,1]$. Equation (3.5) now implies:

$$|F_t(\cup I_i) - L| \le \frac{V(X)L(1 - L)}{2t}. \tag{3.6}$$

Given $\varepsilon > 0$ there exists a set A_ε which includes A and is a disjoint union of intervals in $[0,1]$ such that $m(A_\varepsilon - A) < \varepsilon$ (see Royden, 1968, p.62 and p.39), where $m(A)$ denotes the Lebesgue measure of A. Equation (3.6) leads to

$$|F_t(A) - L| \le |F_t(A) - F_t(A_\varepsilon)| + |F_t(A_\varepsilon) - m(A_\varepsilon)| + |m(A_\varepsilon) - m(A)|$$
$$\le |F_t(A) - F_t(A_\varepsilon)| + \frac{V(X)(L + \varepsilon)(1 - L)}{2t} + \varepsilon.$$

The fact that ε is arbitrary and Proposition 13 in Royden (1968, p.85) imply the desired result. $\qquad\square$

Theorem 3.9.b) *Under the assumptions of Theorem 3.9.a):*

$$\mathrm{d_V}\big((tX)(\mathrm{mod}\,1),\,U\big) \;\leq\; \frac{\mathrm{V}(X)}{8t}.$$

Proof. The result follows directly from the definition of variation distance, Theorem 3.9.a) and the fact that the largest value of $L(1-L)$ is $\frac{1}{4}$.   □

In Theorem 3.9.c) it is shown that if X has bounded variation, then $(tX)(\mathrm{mod}\,1)$ converges to U in the sup norm at a rate at least linear in t^{-1}.

Theorem 3.9.c) *Under the assumptions of Theorem 3.9.a):*

$$\| \,(tX)(\mathrm{mod}\,1) - U \,\|_\infty \;\leq\; \frac{\mathrm{V}(X)}{2t}.$$

Proof. In Theorem 3.9.a) choose $A = [x, x+h]$, divide both sides by h, and let h tend to zero. This leads to

$$|f_t(x) - 1| \;\leq\; \frac{\mathrm{V}}{2t},$$

and the result stated follows. The implicit assumption that $(tX)(\mathrm{mod}\,1)$ has a density follows from Proposition 3.4.   □

Corollary. *Assume X and Y are random variables such that the expectation with respect to Y of the total variation of X given $Y = y$, $\mathrm{E}_Y\,\mathrm{V}(X|Y)$, is finite. Then*

$$\mathrm{d_V}\big((tX + Y)(\mathrm{mod}\,1),\,U\big) \;\leq\; \frac{\mathrm{E}_Y\,\mathrm{V}(X|Y)}{8t}.$$

Proof. Follows from Proposition 2.8.a) and Theorem 3.9.b).   □

The bounds in Theorem 3.9 are sharp, in the sense that there exist random variables for which they are the best possible:

Sharpness of Bounds.   If X is uniform on $[0,1]$ a calculation from first principles shows that

$$f_t(x) = \begin{cases} ([t] + 1)/t, & \text{if } x \leq \{t\}, \\ [t]/t, & \text{if } x > \{t\}, \end{cases}$$

where $[t]$ and $\{t\}$ denote the integer and fractional parts of t.

This implies that

$$\mathrm{d_V}\big((tX)(\mathrm{mod}\,1),\,U\big) = \frac{\{t\}\{1-t\}}{t};$$

$$\|(tX)(\mathrm{mod}\,1) - U\|_\infty = \frac{\max(\{t\}, \{1-t\})}{t}.$$

The total variation of a distribution uniform on $[0,1]$ is equal to 2. Therefore the variation distance bound is attained when $t = k + \frac{1}{2}$, where k is a positive integer.

The sup norm bound may be approximated arbitrarily closely by taking $t = k + \epsilon$, with $\epsilon > 0$ sufficiently small.

Remark 1. Suppose an upper bound for the absolute difference between $\Pr\{0 \leq (tX)(\mathrm{mod}\,1) \leq 10^{-5}\}$ and 10^{-5} is required. The bounds derived in Theorem 3.9.b) for the variation distance between $(tX)(\mathrm{mod}\,1)$ and a distribution uniform on $[0,1]$ are quite inefficient because they do not use the fact that the set under consideration has small Lebesgue measure. Much better bounds can be obtained from Theorem 3.9.a). This theorem implies that

$$\sup_A \left| \frac{\Pr\{(tX)(\mathrm{mod}\,1) \in A\}}{\Pr\{U \in A\}} - 1 \right| \leq \frac{V(X)}{2t},$$

where the supremum is taken over all Borel sets A of positive Lebesgue measure.

Therefore the *relative* error made in approximating $(tX)(\mathrm{mod}\,1)$ by U tends to zero as t tends to infinity.

Remark 2. If $f(x)$ has a kth derivative, $f^{(k)}(x)$, of bounded variation $V_k(X)$, the integration by parts on which the proof of Theorem 3.9.a) is based may be carried out k more times (taking care that at each step to add a suitable constant so that the resulting function integrates to zero), leading to (see Kemperman, 1975):

$$d_V\big((tX)(\mathrm{mod}\,1), U\big) \leq \frac{c_k V_k}{t^{k+1}},$$

where

$$c_k = \begin{cases} 2(1 - 2^{-k-2})|B_{k+2}|\{(k+2)!\}^{-1} & \text{if } k \text{ is even,} \\ |E_{k+1}|\{2^{2k+1}(k+1)!\}^{-1} & \text{if } k \text{ is odd.} \end{cases}$$

The B_j's are the Bernoulli numbers, that is

$$\frac{x}{e^x - 1} = \sum_{i \geq 0} \frac{B_i x^i}{i!}.$$

Therefore $B_0 = 1, B_1 = -\frac{1}{2}, B_2 = \frac{1}{6}, B_{2k+1} = 0$ for $k \geq 1, B_4 = -\frac{1}{30}, B_6 = \frac{1}{42}, B_8 = -\frac{1}{30}, B_{10} = \frac{5}{66}, \ldots$

The E_k's are the Euler numbers, that is

$$\mathrm{sech}x = \sum_{i \geq 0} \frac{E_i x^i}{i!},$$

where $\mathrm{sech}x$ is equal to $2/(e^x + e^{-x})$. Therefore $E_0 = 1, E_2 = -1, E_4 = 5, E_6 = -61, \ldots$

It follows that $(tX)(\mathrm{mod}\,1)$ converges to U at a rate faster than any polynomial rate if X has a density which is differentiable infinitely many times, with all derivatives of bounded variation.

Remark 3. The bounds in the theorem can be easily adapted to the case where $(tX)(\text{mod }\tau)$ is considered instead of $(tX)(\text{mod }1)$:

$$d_V\big((tX)(\text{mod }\tau),\, U[0,\tau]\big) \le \frac{V(X)\tau}{8t},$$

$$\|(tX)(\text{mod }\tau) - U[0,\tau]\,\|_\infty \le \frac{V(X)}{2t}.$$

Remark 4. Denote by $\mathcal{C}$ the set of densities whose derivatives are absolutely integrable. An alternative metric to the variation distance is given by

$$d'_V(X,Y) = \int |f'(x) - g'(x)|dx,$$

with f and g denoting the densities of X and Y. An argument similar to the one used to prove Theorem 3.9.a) shows that

$$d_V\big((tX)(\text{mod }1),(tY)(\text{mod }1)\big) \le \frac{1}{8t}d'_V(X,Y),$$

and hence the function $X \to (tX)(\text{mod }1)$ defined from $(\mathcal{C},d'_V)$ to $(\mathcal{C},d_V)$ is Lipschitz-continuous.

Remark 5. The upper bounds obtained in Theorem 3.9 do not depend on the location of X. It is not the expected value or median of X that determines how fast convergence takes place, but how "spread out" and "wiggly" its density is – as measured by its total variation.

Remark 6. Since $(tX + b)(\text{mod }1) = (t(X + \frac{b}{t}))(\text{mod }1)$ and $V(X) = V(X + \frac{b}{t})$, it follows that

$$d_V\big((tX + b)(\text{mod }1),\, U\big) \le \frac{V(X)}{8t},$$

where b can be any constant. In particular, b could depend on t.

Remark 7. Consider the following density that does not have bounded variation:

$$f(x) = \begin{cases} 1 & \text{if } x \in \cup_{k\ge 1}[k, k + 2^{-k}]; \\ 0 & \text{otherwise.} \end{cases}$$

A calculation from first principles based on (3.2) shows that if $t = 2^k$, k an integer, the density of $(tX)(\text{mod }1)$ satisfies

$$f_t(x) = 1 + \frac{1}{t}\left(\left[\log_2\left(\frac{1}{x}\right)\right] - 1\right),$$

where $[x]$ denotes the integer part of x.

A straightforward calculation now shows that $d_V\big((tX)(\text{mod }1),\, U\big) = 1/2t$. Hence convergence may be linear in t even if X does not have bounded variation and the assumptions of Theorem 3.9.b) do not hold.

It is also interesting to note that in this example – and also when the density of X is not bounded from above – $\|(tX)(\mathrm{mod}\,1) - U\|_\infty = +\infty$, so that there is convergence in the variation distance but not in the sup norm.

3.1.3 Exact Rates of Convergence

If X is a random variable with a density of bounded variation, $(tX)(\mathrm{mod}\,1)$ converges to a distribution uniform on $[0,1]$, U, at a rate at last linear in t^{-1}. If all random variables with densities of bounded variation are considered, this rate cannot be improved upon. Yet if attention is focused on specific random variables, faster rates of convergence may be attained.

Propositions 3.11 gives upper and lower bounds for the variation distance between $(tX)(\mathrm{mod}\,1)$ and U in terms of the Fourier coefficients of X. It is useful in establishing exact rates of convergence in particular cases. The upper bounds are obtained from Poisson's Summation Formula.

Feller (1971, p.63 and p.632) uses Poisson's Summation Formula to study the convergence of $(tX)(\mathrm{mod}\,1)$ to a uniform random variable. Good (1986) establishes upper bounds on the rate of convergence of $(tX)(\mathrm{mod}\,1)$ to a uniform random variable when X is a mixture of normal or Gamma random variables using Poisson's Summation Formula. He also gives a beautiful survey of various statistical applications of Poisson's Summation Formula.

Lemma 3.10 *Let X be a random variable with bounded variation. Denote by $\hat{f}$ the characteristic function of X and by f_t the density of $(tX)(\mathrm{mod}\,1)$. Then:*

$$f_t(x) = \lim_{n \to \infty} \sum_{k=-n}^{n} \hat{f}(2\pi kt)\,e^{-i2\pi kx}.$$

Proof. The expression for $f_t(x)$ follows from Poisson's Summation Formula (for the version being applied here, see Butzer and Nessel, 1971, p.202).

Proposition 3.11 *Let X be a random variable with characteristic function $\hat{f}(t)$. Then*

$$d_V\big((tX)(\mathrm{mod}\,1),\,U\big) \geq \frac{1}{2}\,\sup_{k\in\mathbb{N}} |\hat{f}(2\pi kt)|.$$

Further, if X has bounded variation:

$$d_V\big((tX)(\mathrm{mod}\,1),\,U\big) \leq \sum_{k\geq 1} |\hat{f}(2\pi kt)|.$$

Proof. The lower bound follows from Lemma 3.1 and Proposition 2.7. The upper bound is derived from Lemma 3.10 and Proposition 2.5.b).   □

In the following corollary, exact rates of convergence are established for various well known random variables.

Corollary 1. *If X has a normal density with mean μ and variance σ^2 then*

$$\frac{1}{2}e^{-2\pi^2\sigma^2 t^2} \leq d_V\left((tX)(\bmod 1)\,,\,U\right) \leq e^{-2\pi^2\sigma^2 t^2} + O\left(e^{-8\pi^2\sigma^2 t^2}\right).$$

Corollary 2. *If X has a double exponential density, $f(x) = \frac{1}{2}e^{-|x|}$, then:*

$$\frac{1}{2+8\pi^2 t^2} \leq d_V\left((tX)(\bmod 1)\,,\,U\right) \leq \frac{1}{24t^2}.$$

Corollary 3. *If X has a Cauchy density, $f(x) = 1/\pi(1+x^2)$, then:*

$$\frac{1}{2}e^{-2\pi t} \leq d_V\left((tX)(\bmod 1)\,,\,U\right) \leq e^{-2\pi t} + O\left(e^{-4\pi t}\right).$$

Corollary 4. *If X has a Gamma density with parameters $a > 1, b > 0$, $f(x) = \frac{1}{\Gamma(a)b^a}x^{a-1}e^{-x/b}$; $x > 0$, then:*

$$\frac{1}{2|1+4\pi^2 b^2 t^2|^{a/2}} \leq d_V\left((tX)(\bmod 1)\,,\,U\right) \leq \frac{\sum_{k\geq 1}k^{-a}}{(2\pi b)^a}t^{-a}.$$

Corollary 5. *If X has a density which is a mixture of normal random variables, $f(x) = \int \phi_\theta(x)g(\theta)d\theta$, with $\theta = (\mu,\sigma^2)$, ϕ_θ the corresponding normal random density and $g(\theta)$ a mixing density, then:*

$$d_V\left((tX)(\bmod 1)\,,\,U\right) \leq E_g e^{-2\pi^2\sigma^2 t^2} + O(E_g e^{-8\pi^2\sigma^2 t^2}).$$

In particular, if $\sigma \geq \sigma_0$:

$$d_V\left((tX)(\bmod 1)\,,\,U\right) \leq e^{-2\pi^2\sigma_0^2 t^2} + O(e^{-8\pi^2\sigma_0^2 t^2}).$$

Proof. The results are direct consequences of Proposition 3.11, in the case of part 5 combined with Proposition 3.8.e).   □

Remark. All higher order terms mentioned above can be made explicit and the bounds are therefore not only asymptotic.   □

The fractional part of the sum of independent random variables is now considered. In Proposition 3.12 it is shown that the variation distance to a uniform distribution decreases under convolution.

Assume X_1, X_2, ... are independent, identically distributed random variables with bounded variation and let S_n denote $X_1 + \cdots + X_n$. The variation distance between $S_n(\mathrm{mod}\,1)$ and a distribution uniform on $[0,1]$ tends to zero monotonically at rate c^n, where c is the absolute value of the largest non trivial Fourier coefficient of the X_i's (Proposition 3.13). Kemperman (1975) proved this result using the notion of discrepancy instead of the variation distance. The discrepancy between two distribution functions $F(x)$ and $G(x)$ supported by $[0,1]$ is equal to $\sup_{0 \le x < y \le 1} |(F(x) - F(y)) - (G(x) - G(y))|$.

Proposition 3.12 *Assume X and Y are independent, absolutely continuous random variables and let U denote a distribution uniform on $[0,1]$. Then*

$$d_V\big((X+Y)(\mathrm{mod}\,1),\, U\big) \;\le\; d_V\big(X(\mathrm{mod}\,1),\, U\big).$$

If X has bounded variation and $\widehat{f}_X(t)$ and $\widehat{f}_Y(t)$ denote the characteristic functions of X and Y, respectively, then

$$d_V\big((X+Y)(\mathrm{mod}\,1),\, U\big) \;\le\; \sum_{k \ge 1} |\widehat{f}_X(2\pi k)||\widehat{f}_Y(2\pi k)|.$$

Proof. The first statement follows from conditioning on Y and using Proposition 2.8.a). The second inequality follows from Proposition 3.11. □

Proposition 3.13 (Kemperman) *Let X_1, X_2, ... denote independent, identically distributed random variables with common characteristic function $\widehat{f}(t)$ and denote by S_n the random variable $X_1 + \cdots + X_n$. Define c as the largest among the absolute values of non trivial Fourier coefficients of the X_i's, $c = \sup_{k \ge 1} |\widehat{f}(2\pi k)|$. Assume that the sum of the absolute values of the Fourier coefficients of $\bar{S}_n$ is finite for some $n = n_0$. Let U denote a distribution uniform on the unit interval. Then $S_n(\mathrm{mod}\,1)$ converges in the variation distance to U as n tends to infinity. Furthermore, $d_V(S_n(\mathrm{mod}\,1),\, U)$ tends to zero monotonically. There exists a constant L that only depends on $\widehat{f}(t)$ such that:*

$$\frac{1}{2}c^n \;\le\; d_V(S_n(\mathrm{mod}\,1),\, U) \;\le\; Lc^n, \tag{3.7}$$

and

$$\lim_{n \to +\infty} d_V(S_n(\mathrm{mod}\,1),\, U)^{1/n} \;=\; c. \tag{3.8}$$

Proof. The (trivial) fact that the k-th Fourier coefficient of S_n is equal to $|\widehat{f}(2\pi k)|^n$ and Propositions 3.11 and 3.12 imply that for $n \ge n_0$:

$$\frac{1}{2}c^n \;\le\; d_V(S_n(\mathrm{mod}\,1),\, U) \;\le\; \sum_{k \ge 1} |\widehat{f}(2\pi k)|^n.$$

Letting $L = \sum_{k \geq 1} |\widehat{f}(2\pi k)|^{n_0}/c^{n_0}$ implies (3.7). Equation (3.8) follows from (3.7). Finally, the fact that $d_V(S_n(\mod 1), U)$ decreases monotonically is due to the first part of Proposition 3.12. $\qquad\square$

Remark 1. The assumption that the sum of the absolute values of the Fourier coefficients of S_n is finite for some $n = n_0$ and Billingsley (1986, p.362, problem 26.1) imply that $c < 1$.

Remark 2. The sum of the absolute values of the Fourier coefficients of S_n is finite for some $n = n_0$ if $\widehat{f}(t)$ behaves like t^{-a}, $a > 0$, as t tends to infinity, in particular, if the X_i's have bounded variation.

3.1.4 Fastest Rate of Convergence

What is the fastest possible rate at which $d_V\big((tX)(\mod 1), U\big)$ may tend to zero?

Among the well known densities, the fastest rate is achieved by the normal distribution. This led Aldous, Diaconis and Kemperman to conjecture that the normal density has the fastest rate among all random variables.

The following theorem shows that convergence can be considerably faster than it is for a normal random variable. There exists a family of random variables for which $(tX)(\mod 1)$ is identically uniform from a certain value of t onwards.

Theorem 3.14 *Let X be a random variable with characteristic function $\widehat{f}(t)$. Then $(tX)(\mod 1)$ is uniform on $[0,1]$ for $t \geq t_0$ if and only if $\widehat{f}(t) = 0$ for $t \geq 2\pi t_0$.*

Proof. Let U denote a random variable uniform on $[0,1]$.

The fact that the Fourier coefficients of U are equal to zero, Proposition 2.1 and Lemma 3.1 imply that $(tX)(\mod 1)$ is equal to U if and only if $\widehat{f}(2\pi kt) = 0$, $k = 1, 2, \ldots$ The theorem's conclusion now follows. $\qquad\square$

Remark. There exist many random variables whose characteristic function vanishes from a certain point onwards. Due to Polya's criterion, any continuous function $\widehat{f}(t)$ convex for $t > 0$, satisfying $\widehat{f}(0) = 1$, $\widehat{f}(-t) = \widehat{f}(t)$ and $\lim_{t \to \infty} \widehat{f}(t) = 0$ is a characteristic function. This may be used to construct characteristic functions with compact support.

For example, a random variable with density equal to $f(x) = \frac{1}{2\pi}\left(\frac{\sin(x/2)}{(x/2)}\right)^2$ (see Fig. 3.1) has characteristic function equal to the tent function (see Fig. 3.2). In this case the random variable $(tX)(\mod 1)$ is uniform on $[0,1]$ for $t \geq \frac{1}{2\pi}$.

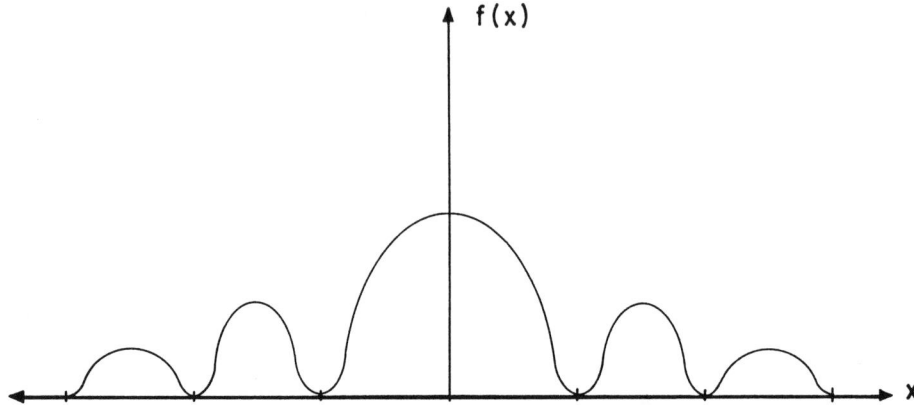

Fig. 3.1. Density that has the tent function as its characteristic function

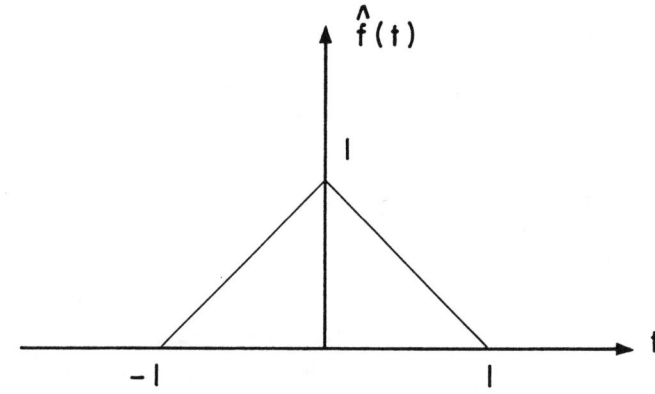

Fig. 3.2. Tent function.

3.2 Applications

3.2.1 A Bouncing Ball

Consider dropping a ball from a height of approximately one foot above a table. The ball will bounce off the table repeatedly (see Fig. 3.3). Assume collisions are perfectly elastic, so that the ball always reaches the same height.

Let $x(t)$ denote the ball's height above the table at time t. Assume the ball starts from rest at a height H. Solving Newton's equations (it is a simple application of the equations of motion for a particle undergoing uniform acceleration) leads to:

$$x(t) = H\{(1 - 4W^2)I_{[0,\frac{1}{2}]}(W) + (1 - 4(1 - W)^2)I_{(\frac{1}{2},1]}(W)\}, \qquad (3.9)$$

with $I_A(x)$ denoting the indicator function of the set A and

$$W = (tS)(\mathrm{mod}\,1), \qquad S = \sqrt{\frac{g}{8H}}. \qquad (3.10)$$

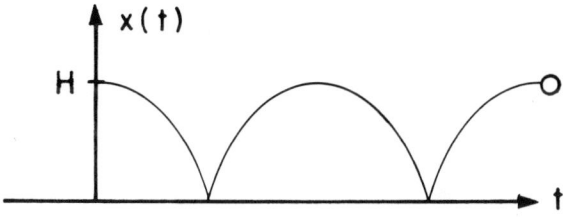

Fig. 3.3. Bouncing ball

If the initial height is known exactly, equation (3.9) completely determines the ball's height at any instant of time. Yet there always is some uncertainty about the ball's initial height. If this uncertainty is modeled as a random variable with a density, $x(t)$ also is a random variable. Its distribution depends on that of H. The set of possible values of $x(t)$ is determined by the possible values of H. Thus, if dependence on initial conditions is to wash away as time passes, the ball's height needs to be standardized. Equation (3.9) motivates looking at $x(t)/H$.

Equation (3.10) implies that $x(t)/H$ is a function of $(tS)(\text{mod}\,1)$ and therefore Theorem 3.2 and the Continuous Mapping Theorem imply that $x(t)/H$ converges, in the weak-star topology, to the corresponding function of a distribution uniform on $[0,1]$. The limiting distribution is independent of the distribution of H due to Theorem 3.3.

Let L denote this limit. The density of L is the following member of the Beta family (see Fig. 3.4):

$$f(l) = \frac{1}{2\sqrt{1-l}}\,; \qquad 0 \le l < 1.$$

The reason why the limiting density is increasing is that the ball travels fastest when it approaches or leaves the table, that is, when $x(t)/H$ is small. Therefore it spends less time near the table than it does near its maximum height H.

The variation distance between $x(t)/H$ and L is bounded by the variation distance between $(tS)(\text{mod}\,1)$ and a uniform random variable due to Proposition 2.6. Applying Theorem 3.9 leads to:

$$d_V\left(\frac{x(t)}{H}, L\right) \le \frac{V(S)}{8t}.$$

To determine how near $x(t)/H$ is from its limiting distribution at a given instant of time, upper bounds for the total variation of the random variable $S = \sqrt{g/8H}$ are needed.

It is natural to think about our uncertainty on the initial height, H, not S. Proposition 3.8.d) is useful to relate both quantities. If the density of H, $f(x)$, tends to zero faster than $x^{-3/2}$ then

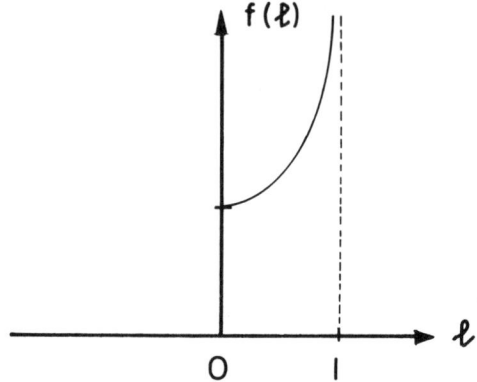

Fig. 3.4. Limiting density for bouncing ball

$$V(H^{-1/2}) \leq 2\int x^{\frac{3}{2}}|f'(x)|dx + 3\int x^{\frac{1}{2}}f(x)dx. \qquad (3.11)$$

If $f(x)$ vanishes for $x \geq L_0$, equation (3.11) and Jensen's inequality imply that:

$$V(H^{-1/2}) \leq 2L_0^{3/2}V(X) + 3\sqrt{EH}.$$

Therefore, for H uniform on 1 foot $\pm$ 2 inches:

$$\mathrm{dv}\left(\frac{x(t)}{H}, L\right) \leq \frac{1.15}{t},$$

and the variation distance will be less than 0.05 after 23 seconds.

Alternatively, if the density of H, $f(x)$, is unimodal, with maximum value M attained at x_0, equation (3.11) implies that

$$V(H^{-1/2}) \leq 4x_0^{3/2}M + 6\sqrt{EH}. \qquad (3.12)$$

Hence, for example, if H is a mixture of normal variables with mean not larger than 14 inches and variance not smaller than 1 inch, equation (3.12) and Proposition 3.8.e) imply that:

$$\mathrm{dv}\left(\frac{x(t)}{H}, L\right) \leq \frac{1.92}{t}.$$

In this case, the largest possible relative error made in approximating the random variable $(tX)(\mathrm{mod}\,1)$ by its limit does not exceed 5 percent after 38.4 seconds.

3.2.2 Coin Tossing

Why do most people believe that the probability of a coin landing heads up is about one half? After all, there exist mechanical coin tossing devices (based on springs) and magicians (with sufficient control over their hands) who are able to toss ten heads in a row.

If initial conditions (position, velocity and angular velocity) are known exactly, there is no uncertainty and which side lands heads up can be determined solving Newton's equations. If the coin is allowed to bounce off the floor, doing the exact physics is not an easy problem: see Yue and Zhang (1985) and Vulovic and Prange (1985) for an analysis of bouncing. Yet, in principle, there is nothing random about a coin flip.

Keller (1986) provides an analysis of coin tossing assuming the coin is caught by the thrower and therefore not allowed to bounce. He shows that as the initial velocity and rate of turning become large, the probability of heads tends to one half. In what follows, Keller's analysis is extended to the case where the coin is allowed to bounce off the floor. Explicit bounds are provided. These are combined with experimental results due to Persi Diaconis to determine how fair actual coin tosses really are.

Consider a circular coin of negligible thickness. Assume that the center of gravity of the coin is at its geometrical center. Suppose that at time $t = 0$, it has an upward velocity v and denote the coin's height and vertical velocity at time t by $y(t)$ and $v(t)$, respectively (see Fig. 3.5).

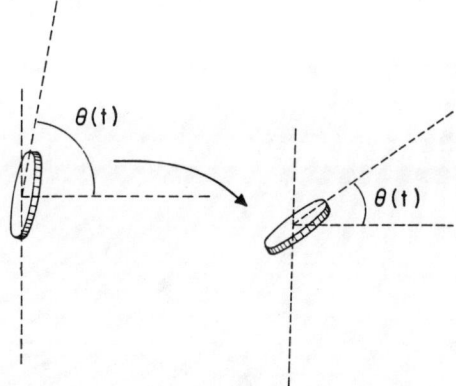

Fig. 3.5. Variables describing a coin's position

In addition to its vertical motion, the coin is assumed to be rotating about a horizontal axis that lies along one of its diameters. Let $\theta(t)$ and $\omega(t)$ denote the angle the coin makes with this axis and its angular velocity at time t, respectively. Assume that initially the coin is horizontal with heads up ($\theta(0) = 0$) and that it has a positive angular velocity ω.

Before considering the case where the coin bounces off the floor, assume it lands on a surface (or is caught in the thrower's hand) when its center of mass reaches the coin's initial height (this simplifies some of the subsequent expressions but is not essential). Also assume that whichever side of the coin is up at that instant remains up.

Let t_1 denote the instant at which the coin is caught. It lands heads up if and only if $(4k - 1)\frac{\pi}{2} \leq \theta(t_1) \leq (4k + 1)\frac{\pi}{2}$ for some integer k.

The equations of motion for a particle subject to uniform acceleration imply that

$$y(t) = vt - \frac{1}{2}gt^2, \tag{3.13}$$

where v denotes the coin's initial linear velocity and g acceleration due to gravity.

The equations for uniform circular motion lead to

$$\theta(t) = \omega t, \tag{3.14}$$

where ω denotes the coin's initial angular velocity.

Equation (3.13) implies that the coin stops at time $t_1 = 2v/g$. This combined with (3.14) shows that the values of the remaining variables describing its motion at the instant it stops are:

$$t_1 = \frac{2v}{g}, \quad \theta(t_1) = \frac{2\omega v}{g}, \quad \omega(t_1) = \omega, \quad y(t_1) = 0, \quad v(t_1) = -v. \tag{3.15}$$

The black region in Fig. 3.6 indicates the combinations of initial vertical and angular velocities for which the coin lands on the surface heads up.

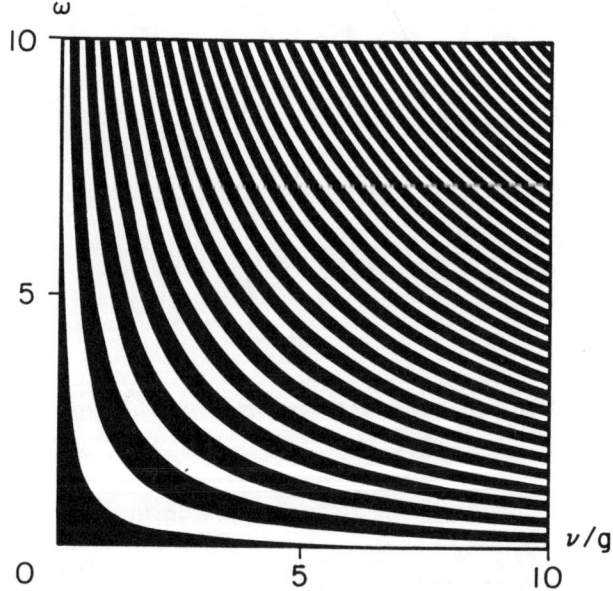

Fig. 3.6. Partition of phase space induced by heads and tails

One way of making mathematically precise the fact that the coin is given a large initial velocity is assuming that the vector of initial velocities, (ω, v), has a joint density whose functional form is fixed, but whose location varies. Keller (1986) shows that the probability of heads tends to one half as this joint density is shifted away from the origin along any ray of the form $\omega = cv$, $c > 0$.

Assume that the initial angular and vertical velocities satisfy $\omega = \omega_0 + a$; $v = v_0 + b$, with $f(\omega, v)$ denoting the joint density of ω_0 and v_0 and $f_v(v)$ and $f_\omega(\omega)$ the corresponding marginals. Suppose the support of $f(\omega, v)$ is included in the first quadrant and let U denote a distribution uniform on $[0, 2\pi]$. Let $V(\omega|v)$ denote the total variation of ω conditioned on the velocity being equal to v and define $V(v|\omega)$ similarly. Then:

$$\left| \Pr\{\text{Heads}\} - \frac{1}{2} \right| \leq \frac{g\pi}{8} \min\left(\frac{V_1}{b}, \frac{V_2}{a} \right), \tag{3.16}$$

where $V_1 = E_v V(\omega|v)$ and $V_2 = E_\omega V(v|\omega)$.

The proof is based on the following string of inequalities:

$$\left| \Pr\{\text{Heads}\} - \frac{1}{2} \right| \leq \text{dv}\left(\frac{2(\omega_0 + a)(v_0 + b)}{g} (\text{mod } 2\pi), U \right)$$

$$\leq \int \text{dv}\left(\left(\frac{2(\omega_0 + a)(v_0 + b)}{g} \middle| v_0 = v \right) (\text{mod } 2\pi), U \right) f_v(v)dv$$

$$\leq \frac{\pi g}{8b} V_1.$$

The first inequality follows from the fact that the coin lands heads up if and only if $\theta(t_1)(\text{mod } 2\pi)$ belongs to the set $[0, \pi/2] \cup [3\pi/2, 2\pi]$, (3.15) and Proposition 2.6; the second from Proposition 2.8.a) and the third from Theorem 3.9.b) $(t = 2(v_0 + g)/b)$, the fact that $v_0 \geq 0$ and Proposition 3.8.a). A similar argument leads to the second inequality needed to establish (3.16). This argument shows that if either velocity or rate of spin is large, the outcome becomes random.

It is now shown that the bounds obtained in (3.16) remain valid if the coin is allowed to bounce. To deal with this case, attention is shifted to the instant t_1 when the coin begins to fall to the floor instead of being caught in the thrower's hand. Equation (3.15) implies that $\omega(t_1) = -g\theta(t_1)/2v(t_1)$. Hence $\theta(t_1)$ and $v(t_1)$ determine the coin's final position and there exists a function $h(\theta, v)$ which associates to every combination of $\theta(t_1)$ and $v(t_1)$ the corresponding outcome: either heads or tails.

Since θ is an angular variable, $h(\theta, v)$ has period 2π in its first argument.

Assume that for a given set of initial conditions the coin lands heads up. If the coin is allowed to bounce again, but this time its initial angular position is modified by 180 degrees, the remaining conditions unchanged, it is natural to expect that this time it will land tails up. That is, $h(\theta, v)$ equals heads if and only if $h(\theta + \pi, v)$ equals tails. This is a basic symmetry assumption. It is also present in the analysis of bouncing undertaken by Yue and Zhang (1985) and Vulovic and Prange (1985). Since these authors do the actual physics, they need additional assumptions. No additional assumptions on how the coin bounces are needed here.

The symmetry condition implies that

$$\Pr\{h(U, v) = \text{Heads}\} = \frac{1}{2}, \tag{3.17}$$

where U is independent of v. Equation (3.17) and Proposition 2.6 lead to

$$\left| \Pr\{\text{Heads}\} - \frac{1}{2} \right| = \left| \Pr\{h(\theta, v) = \text{Heads}\} - \frac{1}{2} \right|$$

$$= | \Pr\{h(\theta(\text{mod } 2\pi), v) = \text{Heads}\} - \Pr\{h(U, v) = \text{Heads}\}|$$

$$\leq \text{d}_{\text{V}}\left((\theta(t_1)(\text{mod } 2\pi), v), (U, v)\right),$$

and Proposition 2.9 reduces the proof to the final step in the derivation of (3.16).

What are typical values for the initial linear and angular velocities in actual coin tosses?

A simple calculation based on the equations of motion for a particle undergoing uniform acceleration shows that the coin's initial velocity is approximately 8 ft/sec. for flips that reach a height of one foot. Persi Diaconis devised an ingenious method – based on tying dental floss to a coin, tossing the coin and then carefully unwinding it – to obtain estimates for the initial angular velocity, ω, of the coin. He concluded that typical values for ω were centered at 38 rev./sec. For example, if v and ω are assumed independent uniform between 7 and 9 feet/sec and 36 and 40 rev./sec, respectively, equation (3.16) implies that

$$\left| \Pr\{\text{Heads}\} - \frac{1}{2} \right| \leq 0.056. \tag{3.18}$$

The spread of the joint density describing the coin's initial linear and angular velocities used above are conservative. For most people, the density on initial conditions is more spread out and the bounds are sharper. It is suggested that the reader flip a coin many times to check this statement.

Proposition 3.11 shows that rates of convergence can be much faster if the random variables involved are normal. If the conditional distributions of v given ω are all mixtures of normal variables with standard deviations larger than σ, and ω takes values larger than a, a similar calculation to the one leading to (3.16), but using Corollary 5 of Proposition 3.11 instead of Theorem 3.9 at the final step, leads to

$$\left| \Pr\{\text{Heads}\} - \frac{1}{2} \right| \leq e^{-2\sigma^2 a^2/g^2}. \tag{3.19}$$

A natural point of comparison with (3.18) is obtained by letting $\sigma = \frac{1}{2}$ ft./sec. and $a = 72\pi$ rad./sec. Equation (3.19) then becomes:

$$\left| \Pr\{\text{Heads}\} - \frac{1}{2} \right| \leq 1.5 \times 10^{-11}.$$

3.2.3 Throwing a Dart at a Wall

The following example is from Diaconis and Engel (1986). Consider throwing a real dart at a real wall. If the left half of the wall is painted black, and the right half painted white, there is nothing very random about the outcome: by aiming a bit to the left, the dart winds up in the black section.

Now suppose the paint is rearranged to form stripes which are alternately painted black and white. If the distance between the stripes is large, things still aren't random, but as the distance gets smaller, black and white will be judged nearly equally likely by almost anyone (see Fig. 3.7).

Fig. 3.7. Wall painted with stripes

It is not difficult to give quite sharp quantitative estimates: Suppose the thrower stands a distance l from the wall, and the stripes have width d. Clearly only the ratio l/d matters, so without loss of generality, take $l = 1$.

Let θ be the angle of release of the dart and suppose $f(\theta)$ is a probability density on $(0, \pi)$. If θ is chosen from f,

$$\Pr\{\text{Black}\} = \sum_{n=-\infty}^{\infty} \Pr\{\cot^{-1}(2n+1)d \leq \theta \leq \cot^{-1}(2nd)\}$$

$$= \Pr\left\{ \left(\frac{\cot \theta}{2d} \right) (\bmod\, 1) \leq \frac{1}{2} \right\},$$

where $x(\bmod\, 1)$ denotes the fractional part of the real number x (see Fig. 3.8).

Fig. 3.8. Relation between stripes and angle of release

Assuming θ has a continuously differentiable density $f(\theta)$ of bounded variation $V(\theta)$, Theorem 3.9 and Proposition 3.8.d) imply that

$$\left| \Pr\{\text{Black}\} - \frac{1}{2} \right| \leq \frac{V(\theta) + 1}{4} d.$$

How can one arrive at concrete estimates? Even if the thrower aims at black, the random variable describing her angle of release takes values over a set which cannot be arbitrarily small. Suppose that, conditional on the region she aims at, the range of possible values of θ is at least 10 degrees. Also assume that within this range, the conditional density describing θ does not oscillate too wildly. This can be made mathematically precise by requiring that its maximum value be bounded, say, by twice the value of the corresponding uniform random variable.

If the thrower is trying to maximize her probability of hitting black, she is aiming at the center of one of the black sections and therefore her density, conditional on the black section she has chosen, is unimodal.

The argument given above leads to a density for θ (measured in radians) that is a mixture of unimodal densities bounded by $72/\pi$. Proposition 3.8.e) implies that

$$\left| \Pr\{\text{Black}\} - \frac{1}{2} \right| \leq 6\,d.$$

For arbitrary l this gives:

$$\left| \Pr\{\text{Black}\} - \frac{1}{2} \right| \leq \frac{6d}{l}.$$

Assume the thrower is ten feet from the wall. No matter how hard she aims at a black region, her probability of hitting black is within 0.05 of one half if the width of the stripes does not exceed one inch. The width of the stripes has to be smaller than one fiftieth of an inch to ensure that the probability of hitting black be within 0.001 of one half.

If the distance l between the thrower and the wall is not known exactly, a similar argument and Proposition 2.8.a) lead to:

$$\left| \Pr\{\text{Black}\} - \frac{1}{2} \right| \leq \mathrm{E}\left(\frac{V(\theta| l) + 1}{4\,l} \right) d,$$

where the expectation is taken with respect to l and $V(\theta| l)$ denotes the total variation of θ conditioned on the value of l.

If $l \geq l_0$:

$$\left| \Pr\{\text{Black}\} - \frac{1}{2} \right| \leq \frac{\mathrm{E}_l V(\theta| l) + 1}{4\,l_0}\, d.$$

This bound decreases as l_0 increases, and is still linear in d.

3.2.4 Poincaré's Roulette Argument

Poincaré's most famous application of the method of arbitrary functions is his analysis of roulette (see Poincaré, 1896).

He considers a wheel which is divided into a large number, n, of circular sections of equal size painted alternatingly black and white (see Fig. 3.9). A large initial impulse is given to the wheel. Attention is focused on a point which remains fixed while the wheel rotates. When the wheel stops, one of its sections will coincide with the fixed point. What is the probability that this section is black?

Fig. 3.9. Black and white sections on roulette wheel

Poincaré does not say how the wheel stops for he is not going to do the physics. He argues that in general the wheel's final position is described by a density and that as the number of white and black sections grows, the area under this density corresponding to white sections approaches one half.

Let X denote a random variable describing the wheel's final position with respect to the fixed point. By conveniently choosing coordinates, the event that the fixed point ends on a black section is equal to $\{X \in [0, \frac{2\pi}{n}] \cup [\frac{4\pi}{n}, \frac{6\pi}{n}] \cup \ldots\}$.

Therefore:

$$\Pr\{\text{Black}\} = \Pr\left\{\left(\frac{nX}{4\pi}\right)(\text{mod}\,1) \leq \frac{1}{2}\right\}.$$

Assume the characteristic function of X vanishes at infinity (in particular, that X has a density). Then the probability of a black section landing on the fixed point tends to one half as n grows (see Theorem 3.2). Theorem 3.9 gives concrete upper bounds on how fast convergence takes place by making the additional assumption that X has a density of bounded variation $V(X)$:

$$\left|\Pr\{\text{Black}\} - \frac{1}{2}\right| \leq \frac{\pi V(X)}{2n}.$$

3.2.5 Poincaré's Law of Small Planets

Poincaré (1896, p.129ff and 1952, p.196ff) uses the method of arbitrary functions to determine the distribution of minor planets on the zodiac as follows.

Since the number of planets is large, the empirical distribution function of their longitudes at any instant in time – in particular at time $t = 0$ in the distant past – may be approximated by a random variable X with a continuous density. Similarly, a random variable Y with a continuous density can be used to describe the distribution of planets' angular velocities – which does not change over time. It follows that the random variable

$$L(t) = (tX + Y)(\mod 2\pi),$$

describes the distribution of planets' longitudes at time t.

Assuming the joint density of X and Y is sufficiently smooth (continuous, differentiable with respect to x) Poincaré (1896, p.129ff) uses Fourier methods to prove that $(tX + Y)(\mod 2\pi)$ converges in the weak-star topology to a random variable uniform on $[0, 2\pi]$, U, as t tends to infinity. He concludes that *"after enough time has passed, the planets are distributed uniformly among all signs of the zodiac."*

Theorem 3.3 implies that $L(t)$ converges in the weak-star topology to U for any joint density describing X and Y. Further, the limiting random variable U is independent of the planet's initial position and velocity. The Corollary of Theorem 3.9 gives upper bounds for the variation distance between $(tX + Y)(\mod 2\pi)$ and U.

It is interesting to note that – as mentioned above – Poincaré interprets the density on initial conditions as a smooth approximation of the empirical distribution function.

3.2.6 An Example from the Dynamical Systems Literature

An area of research that has been very active in the last two decades is that of dynamical systems. Mathematical models have been constructed which isolate simplifying features of dissipative dynamical systems, i.e. of physical systems with some sort of "friction." Such systems tend in general to equilibrium positions when left alone. But when they are driven from the outside, they quite often show behavior which is called erratic, aperiodic, turbulent, or strange.

Iterations of continuous maps of an interval into itself serve as the simplest example of models for dynamical systems. For an excellent introduction and review of this topic see Collet and Eckmann (1980). It is assumed that the system's state space is the unit interval and that it is observed at discrete intervals of time. Its evolution is described by a fixed function $g(x)$: if the system's position at time n is x_n then its position at time $n + 1$ is $g(x_n)$, with $g(x)$ fixed.

One of the main objectives of the dynamical systems literature is to make mathematically precise the idea of "unpredictability" of a physical system. Many concepts have been used to this effect: sensitivity to initial conditions, Lyapunov exponents, etc.

Consider the following example. Let $g(x)$ be the tent function (see Fig. 3.10):

$$g(x) = 1 - 2\left|x - \frac{1}{2}\right|.$$

The system starts off at time zero from a point described by a random variable X_0 (which takes values in $[0,1]$). At time n its position is $X_n = g^n(X_0)$, where g^n denotes the n-fold composition of g. In this particular example, g^n corresponds to the saw-tooth function (see Figure 3.11).

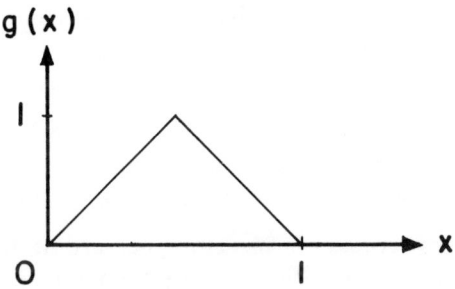

Fig. 3.10. Tent function $g(x)$

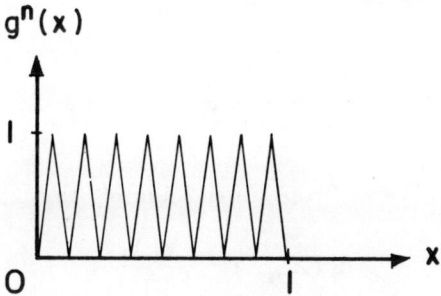

Fig. 3.11. Saw-tooth function

It is now shown that X_n converges, in the variation distance, to a distribution uniform on the unit interval if X has bounded variation $V(X)$.

It is easy to see that

$$\{X_n \leq u\} = \{(2^n X_0)(\bmod 1) \in [0, \tfrac{u}{2}] \cup [1 - \tfrac{u}{2}, 1]\}. \tag{3.20}$$

Equation (3.20) and Theorem 3.9.a) imply that

$$\left| \frac{1}{h} \Pr\{u \leq X_n \leq u + h\} - 1 \right| \leq \frac{V(X)}{2^n},$$

and therefore (let h tend to zero and use Proposition 2.5.b) :

$$d_V(X_n, U) \leq \frac{V(X_0)}{2^{n+1}} = \frac{V(X_0)}{2} e^{-n \log 2}. \tag{3.21}$$

Theorem 5.2 (see Chap. 5) implies that X_n converges to U in the variation distance for any absolutely continuous random variable X_0 describing initial conditions. Yet bounds on the rate of convergence like those derived in (3.21) require additional smoothness assumptions on the density of X.

The method of arbitrary functions makes the idea of "unpredictability" precise by stating that, for large n, the distribution of X_n is near a limiting distribution which does not depend on the density describing the initial state X_0. Only trivial predictions of values in the distant future are possible, and in this sense the system is unpredictable.

A function's Lyapunov exponents measure the rate at which two nearby points separate. In the case of the tent function its only Lyapunov exponent is equal to $\log 2$. Equation (3.21) therefore shows that the rate at which X_n converges to a uniform random variable is related to the corresponding Lyapunov exponent.

Ulam and von Neumann (1947) considered iterations under the map $g(x) = 4x(1-x)$ (strictly speaking they worked with the equivalent map on $[-1,1]$). Making a change of variables (see Collet and Eckmann, 1980, p.15) reduces this system to the tent map. From the corresponding change of variables it follows that the limiting distribution, S, is an arcsine law with density

$$ f(s) \; = \; \frac{1}{\pi \sqrt{s(1-s)}} \; , \qquad 0 < s < 1. $$

Further, equation (3.21) and Proposition 2.6 imply that

$$ d_V(X_n, S) \; \leq \; \frac{V(X_0)}{2^{n+1}} \, . $$

Another famous example that can be analyzed along the same lines – see Whittle (1983, p.24) for a discussion of its importance in time series analysis – is the one dimensional baker transformation, $X_{n+1} = (2X_n)(\bmod 1)$. In principle, any iterated map of the unit interval may be analyzed from the point of view of the method of arbitrary functions. Yet the mathematics developed in this chapter only applies to the particular cases – such as those mentioned above – where the mappings may be expressed in terms of $(tX)(\bmod 1)$.

4. Higher Dimensions

In this chapter the results of Chap. 3 are extended to physical systems with many degrees of freedom. Given a random vector $X = (X_1, \ldots, X_n)'$, the behavior of $(tX)(\mathrm{mod}\,1) = ((tX_1)'(\mathrm{mod}\,1), \ldots, (tX_n)(\mathrm{mod}\,1))$ is studied in detail in Sect. 4.1. A necessary and sufficient condition for weak-star convergence of $(tX)(\mathrm{mod}\,1)$ to a distribution uniform on $[0, 1]^n$, U_n, as t tends to infinity, is established in Theorem 4.2 (Borel, Hopf, Kallenberg). The random vector $(tX)(\mathrm{mod}\,1)$ converges to U_n in the variation distance if and only if X has a density (Theorem 5.3).

The concept of bounded variation is extended to higher dimensions (Sect. 4.1.2) leading to the notion of bounded mean-conditional variation. Upper bounds for the variation distance between $(tX)(\mathrm{mod}\,1)$ and U_n that are linear in t^{-1} are derived for random vectors satisfying this condition (Theorem 4.9). They apply, in particular, if the density of X has integrable partial derivatives.

Convergence can take place at rates faster than linear in t^{-1} if specific families of distributions are considered (Proposition 4.14). Further, there exists a family of random vectors for which the distribution of $(tX)(\mathrm{mod}\,1)$ is identically uniform once t passes a certain threshold (Proposition 4.15).

The results of Sect. 4.1 are applied to various examples in Sect. 4.2. They include a heavy symmetric top (as a particular case of an integrable system), a coupled harmonic oscillator (as an example of a small oscillations problem), billiards, gas molecules in a room, an application to random number generators based on physical devices and repeated observations of various examples considered in Chap. 3. These applications may be read independently from the section containing the mathematical results.

4.1 Mathematical Results

4.1.1 Weak-star Convergence

The main result of this subsection is Theorem 4.2 which provides necessary and sufficient conditions for weak-star convergence of $(tX)(\mathrm{mod}\,1)$ to U_n, as t tends to infinity. Its proof is a straightforward generalization of the one dimensional case. This result may be traced back to Borel (1909) when X has a uniform distribution on a rectangle included in the unit square. Hopf (1934) proves it for any absolutely continuous random vector. Kallenberg (1980, Theorems 4.1 and 4.3) has a proof of the general result.

Lemma 4.1 *Let X denote a real valued, n dimensional, random vector with characteristic function $\hat{f}_X(\lambda)$, $\lambda \in \mathbb{R}^n$. Denote $X(\mathrm{mod}\,1)$ by Y and the corresponding characteristic function by $\hat{f}_Y(\lambda)$. Then the Fourier coefficients of Y are equal to those of X, that is*

$$\hat{f}_Y(2\pi m) = \hat{f}_X(2\pi m); \qquad m \in \mathbb{Z}^n.$$

Proof. Straightforward generalization of the proof of Lemma 3.1.   □

In the one dimensional case, a necessary and sufficient condition for weak-star convergence of $(tX)(\mathrm{mod}\,1)$ to a distribution uniform on the unit interval is that X satisfy the Riemann-Lebesgue condition, that is, that its characteristic function vanish at infinity. The corresponding generalization to higher dimensions requires the following definition:

Definition. An n dimensional random vector X satisfies the Riemann-Lebesgue condition if $m \cdot X$ satisfies the (one dimensional) Riemann-Lebesgue condition for every n dimensional integer valued vector m different from the null vector, that is, if

$$(\forall m \in \mathbb{Z}^n_*) \quad \lim_{t \to \infty} \hat{f}(tm) = 0,$$

where $u \cdot v$ denotes the usual inner product in $\mathbb{R}^n$.

Theorem 4.2 (Borel, Hopf, Kallenberg) *Assume X is an n dimensional random vector and t a real number. A necessary and sufficient condition for weak-star convergence of $(tX)(\mathrm{mod}\,1)$ to U_n, as t tends to infinity, is that X satisfy the Riemann-Lebesgue condition.*

Proof. Follows directly from Lemma 4.1 and Proposition 2.1.   □

Corollary 1. *If X is an n dimensional random vector with a density (with respect to Lebesgue measure) then $(tX)(\mathrm{mod}\,1)$ converges to U_n, in the weak-star topology, as t tends to infinity.*

Proof. This is a consequence of the n dimensional version of the Riemann-Lebesgue Lemma (see Stein and Weiss, 1971, p.2).   □

Corollary 2. *Assume X and Y are n dimensional random vectors and the conditional distribution of X given $Y = y$ is well defined and satisfies the Riemann-Lebesgue condition for y in a set to which Y assigns probability one. This holds, in particular, if (X, Y) is absolutely continuous. Then, as t tends to infinity, $(tX + Y)(\mathrm{mod}\,1)$ converges to U_n in the weak-star topology.*

Proof. Similar to that of Corollary 2 of Theorem 3.2.   □

Remark. The generality gained by working with distributions which satisfy the Riemann-Lebesgue condition instead of considering absolutely random vectors might seem of little interest. Yet there are occasions on which a function of an absolutely continuous random variable does not have a density (because its support is of lower dimension) but still satisfies the Riemann-Lebesgue condition.

For example, let X be a random variable satisfying the (one dimensional) Riemann-Lebesgue condition and assume $(tX)(\bmod 1)$ is observed at various instants in time. The random variables resulting from these observations are highly dependent because they are all a function of X. It is then interesting to determine whether it possible to choose the instants of time at which $(tX)(\bmod 1)$ is observed in such a way that the observations approach independence as time passes.

Let $Y(t) = (tX, \alpha tX)(\bmod 1)$, where α is a fixed real number. Denote the characteristic function of X by $\widehat{f}(\lambda)$. The characteristic function of the random variable $m_1 X + m_2 \alpha X$, evaluated at λ, is equal to $\widehat{f}((m_1 + m_2 \alpha)\lambda)$ and therefore Theorem 4.2 implies that $Y(t)$ converges in the weak-star topology to a distribution uniform on the unit square if and only if α is irrational.

Note that, in this example, not only does the random vector $(X, \alpha X)$ not have a density but its characteristic function does not vanish at infinity either: along the ray of slope α it is equal to one. Therefore absolute continuity of a random variable is strictly stronger than having its characteristic function vanish at infinity which in turn is strictly stronger than having it satisfy the Riemann-Lebesgue condition.   □

The previous example is a particular case of the following result.

Proposition 4.3 *Assume X is an n dimensional random vector with characteristic function $\widehat{f}(\lambda)$ vanishing at infinity. This is the case, in particular, if X has a density. Let $\tau_1(t), \ldots, \tau_r(t)$ denote real valued functions of a real variable and define*

$$Y(t) = \big(\tau_1(t)X, \ldots, \tau_r(t)X\big)(\bmod 1).$$

That is, $Y(t)$ is the process that results from observing $(tX)(\bmod 1)$ at instants of time determined by the $\tau_i(t)$'s.

The random vector $Y(t)$ converges in the weak-star topology to a distribution uniform on $[0, 1]^{nr}$, U_{nr}, as t tends to infinity, if

$$(\forall m \in \mathbb{Z}_*^{nr}) \ \lim_{t \to +\infty} \Big\| \sum_i m_i \tau_i(t) \Big\| = +\infty, \tag{4.1}$$

where $\|\cdot\|$ denotes any norm in $\mathbb{R}^n$ and $m = (m_1, \ldots, m_r)$ is an integer valued vector in $\mathbb{Z}^{nr}$ with the m_i's in $\mathbb{Z}^n$ and at least one of them having a coordinate different from zero.

If the characteristic function of X is not equal to zero for any $\lambda \in \mathbb{R}^n$, the condition established in (4.1) is also necessary.

Proof. It is a particular case of Proposition 6.5 (that is proved in Chap. 6).   □

Example 1. Assume the process $(tX)(\mathrm{mod}\,1)$ is observed at instants $\alpha_1 t,\ldots,\alpha_n t$. How should $\alpha_1,\ldots,a_n$ be chosen so that $Y(t)$ is approximately uniform for large values of t?

Proposition 4.3 shows that $Y(t)$ converges to a distribution uniform on $[0,1]^{nr}$ if the α_i's are linearly independent over the field of the rational numbers, that is, if there do not exist integers $p_1,\ldots,p_n$ not all equal to zero, such that $\sum p_i \alpha_i = 0$. That this condition is also necessary follows from the fact that otherwise there exist non trivial Fourier coefficients of $Y(t)$ converging to one as t tends to infinity.

If $(tX)(\mathrm{mod}\,1)$ is observed at instants that are equally spaced on a logarithmic scale, that is, $\tau_k(t) = e^{k+t}$, $k = 1,\ldots,r$, then $Y(t)$ is approximately uniform for large t. Further, the argument given in the proof of Proposition 6.13 can be used to show that the infinite dimensional stochastic process

$$Z(t) \;=\; \bigl(\tau_1(t)X, \tau_2(t)X,\; \ldots\; \bigr)(\mathrm{mod}\,1)$$

converges weak-star (with respect to the topology generated by the cylinder sets) to a process uniform on $[0,1]^{\infty}$.

Example 2. Assume $(tX)(\mathrm{mod}\,1)$ is observed at instants $c_1 t, c_2 t^2,\ldots,c_r t^r$ where $c_1,\ldots,$ c_n are different from zero. Proposition 4.3 implies that the resulting random vector converges, in the weak-star topology, to a distribution uniform on $[0,1]^{nr}$ as t tends to infinity.

Example 3. In the previous examples the instants at which $(tX)(\mathrm{mod}\,1)$ is observed grow apart as t tends to infinity. To see that this condition is necessary for convergence to a uniform distribution consider

$$Y(t) \;=\; \bigl(tX, (t+k)X\bigr)(\mathrm{mod}\,1).$$

From the fact that

$$\bigl\{\bigl((t+k)X\bigr)(\mathrm{mod}\,1) - (tX)(\mathrm{mod}\,1)\bigr\}(\mathrm{mod}\,1) \;=\; (kX)(\mathrm{mod}\,1),$$

it is clear that dependence on X does not wash away as t tends to infinity. Using Lemma 4.1 to calculate the Fourier coefficients of $Y(t)$ and applying Proposition 2.1 shows that $Y(t)$ does have a limit as t tends to infinity and that this limit depends on the distribution of X. $\qquad\square$

For random vectors satisfying the conditions of Theorem 4.2, $(tX)(\mathrm{mod}\,1)$ does not depend much on the initial distribution of X when t is large. One might therefore expect that X and $(tX)(\mathrm{mod}\,1)$ become approximately independent as t grows. This idea is made precise in the following theorem, which shows that the random vector $(tX)(\mathrm{mod}\,1)$ becomes independent of almost any fixed distribution as t tends to infinity.

Theorem 4.4 (Hopf) *Assume X and Y are n dimensional random vectors and Z is a p dimensional random vector such that (X,Y,Z) is absolutely continuous. Then, as t tends to infinity, $(tX + Y)(\mathrm{mod}\,1)$ converges in the weak-star topology to a distribution uniform on $[0,1]^n$ that is independent of (X,Y,Z).*

Proof. It is first shown that, given $x_0, y_0 \in \mathbb{R}^n$ and $u = (u_1, \ldots, u_n) \in [0,1]^n$,

$$\lim_{t \to \infty} \Pr\{X \leq x_0, Y \leq y_0, (tX + Y)(\bmod 1) \leq u_0\} = \Pr\{X \leq x_0, Y \leq y_0\} \Pi_{i=1}^n u_i, \quad (4.2)$$

where inequalities involving vectors are interpreted coordinatewise. This implies that the limiting distribution is independent of (X, Y).

Consider the identity

$$\Pr\{X \leq x_0, Y \leq y_0, (tX + Y)(\bmod 1) \leq u_0\}$$
$$= \Pr\{X \leq x_0, Y \leq y_0\} \Pr\{(tV + W)(\bmod 1) \leq u_0\},$$

where (V, W) denotes $((X|X \leq x_0, Y \leq y_0), (Y|X \leq x_0, Y \leq y_0))$.

The random vector (V,W) has a density (which is proportional to that of (X, Y) when $(x, y) \leq (x_0, y_0)$ and zero otherwise). Hence Corollary 2 of Theorem 4.2 leads to (4.2).

To show that (X, Y, Z) is independent of the limiting distribution, apply (4.2) to $(X|Z \leq z_0)$ and $(Y|Z \leq z_0)$ in the identity

$$\Pr\{X \leq x_0, Y \leq y_0, Z \leq z_0, (tX + Y)(\bmod 1) \leq u_0\} =$$
$$= \Pr\{X \leq x_0, Y \leq y_0 \ (tX + Y)(\bmod 1) \leq u_0 | Z \leq z_0\} \Pr\{Z \leq z_0\},$$

and let t tend to infinity. This completes the proof. $\qquad\qquad\square$

4.1.2 Bounds on the Rate of Convergence

The main result in this subsection is Theorem 4.9 that provides tractable upper bounds for the variation distance between $(tX)(\bmod 1)$ and a distribution uniform on $[0,1]^n$.

As pointed out in Chap. 3, assumptions stronger than those of Theorem 4.2 are needed to ensure good rates of convergence.

In the one dimensional case, the total variation of a random variable played an important role when obtaining bounds on the rate of convergence: the bounds were directly proportional to it. The corresponding concept in higher dimensions is introduced in the following definition.

Definition. Assume X is a random variable and Y a random vector such that the conditional distribution of X given $Y = y$ has a density (on a set of probability one with respect to Y). The *mean-conditional variation of X given Y*, $V_1(X, Y)$, is defined as the expected value (with respect to Y) of the total variation of X given Y, that is,

$$V_1(X, Y) = E_Y V(X|Y) = \int V(X|Y = y) dF_Y(y), \quad (4.3)$$

where $F_Y(y)$ denotes the distribution function of Y.

The random variable X is said to have bounded mean-conditional variation with respect to Y if $V_1(X, Y)$ is finite.

Remark. For alternative extensions of the concept of total variation to higher dimensions, useful in quasi Montecarlo methods, see Kuipers and Niederreiter (1974, p.147ff).   □

The main properties of the notion of mean-conditional variation are given in the following two propositions:

Proposition 4.5 *Assume the random variable X has bounded mean-conditional varia-tion with respect to the n dimensional random vector Y. Denote the density of X given $Y = y$ by $f(x|y)$ and the distribution of Y by $F_Y(y)$. Then:*

a)   *For any real number $t \neq 0$, $V_1(tX, Y) = V_1(X, Y)/|t|$.*

b)   *For any real number a and any n dimensional vector b:*

$$V_1(X + a, Y + b) = V_1(X, Y).$$

c)   *If $x \to f(x|y)$ is unimodal, with maximum value $m(y)$, then:*

$$V_1(X, Y) = 2 \int m(y) dF_Y(y).$$

d)   *If $x \to f(x|y)$ is piecewise continuously differentiable with jumps $\delta_1(y), \ldots, \delta_{n(y)}(y)$, then*

$$V_1(X, Y) = \int\int \left| \frac{\partial f}{\partial x}(x, y) \right| dx \, dF_Y(y) + \int \sum_{i=1}^{n(y)} \delta_i(y) dF_Y(y).$$

Now suppose (X, Y) has a density, $f(x, y)$, and denote the support of Y by $\mathcal{Y}$.

e)   *Let $V\big(f(\cdot, y)\big)$ denote the total variation of the function $x \to f(x, y)$. Then:*

$$V_1(X, Y) = \int_{\mathcal{Y}} V\big(f(\cdot, y)\big) dy.$$

Proof. **a)**   Follows from the definition of mean-conditional variation and Proposition 3.8.a).

b) Proposition 3.8.a) and a change of variable imply that

$$\begin{aligned}
V_1(X + a, Y + b) &= E_{Y+b} V_1(X + a \,|\, Y + b) \\
&= \int V(X + a \,|\, Y + b) dF_{Y+b}(y) \\
&= \int V(X \,|\, Y + b = y) dF_{Y+b}(y) \\
&= \int V(X \,|\, Y = u) dF_Y(u) \\
&= V_1(X, Y).
\end{aligned}$$

c) Follows from the definition of mean-conditional variation and Proposition 3.8.c).

d) Follows from the definition of mean-conditional variation and Proposition 3.8.b).

e) As $f_{X|Y}(x|y) = f_{X,Y}(x,y)/f_Y(y)$:

$$V_1(X,Y) = E_Y V(X|Y)$$

$$= \int_y V\big(f_{X|Y}(\cdot\,|y)\big) f_Y(y)dy$$

$$= \int_y V\big(f(\cdot,y)\big) dy. \square$$

Example 1. Assume (X,Y) has a bivariate normal distribution and denote the variance of X and correlation between X and Y by σ_X^2 and r, respectively. The fact that the conditional distribution of X given $Y = y$ is normal with variance $\sigma_X^2(1 - r^2)$ and Proposition 4.5.c) imply that $V_1(X,Y) = 2/\sigma_X \sqrt{2\pi(1 - r^2)}$. Therefore, other things equal, the conditional variation variation of X given Y is smallest when X and Y are independent (that is, when $r = 0$) and it tends to infinity as X and Y become more and more correlated.

Example 2. Assume (X,Y) has a uniform distribution on a convex open subset $\mathcal{C}$ of $\mathbb{R}^{p+1}$, with X a random variable and Y a p dimensional random vector. Denote the support of Y, this is, the corresponding p dimensional projection of $\mathcal{C}$, by $\mathcal{Y}$. Proposition 4.3.b) implies that the conditional variation of X given Y is equal to $2\mathrm{Vol}(\mathcal{Y})/\mathrm{Vol}(\mathcal{C})$, where $\mathrm{Vol}(\cdot)$ denotes the corresponding p or $p + 1$ dimensional volume.

For example, if (X,Y) has a distribution uniform on a (two dimensional) disc of radius R, $V_1(X,Y)$ is equal to $4/\pi R$. $\qquad\qquad\square$

In Example 1 the total variation of X given Y is smallest when X and Y are independent. The following proposition shows that this holds quite generally:

Proposition 4.6 *Assume the random vector (X,Y,Z) has a density, where X is a random variable and Y and Z random vectors. Then*

$$V_1(X,Y) \leq V_1\big(X,(Y,Z)\big). \qquad (4.4)$$

Therefore, in particular:

$$V(X) \leq V_1(X,Y). \qquad (4.5)$$

Hence the mean-conditional variation of a random variable is always at least as large as its total variation.

A sufficient condition for equality in (4.5) is that X and Y be independent. There are occasions, though, where equality holds in (4.5) and X and Y are dependent.

Proof. Here – and in many other proofs throughout this chapter – the letter f is used to denote all densities involved. The argument of f indicates which random variable is being considered. For example, $f(x,y)$ denotes the density of (X,Y) while $f(x,y|z)$ denotes the density of (X,Y) conditional on $Z = z$.

Consider arbitrary real numbers $x_0, \ldots, x_n$ with $x_0 < \ldots < x_n$. The fact that

$$f(x,y) = \int f(x,y,z)dz$$

implies that

$$\sum_{k=1}^{n} |f(x_k, y) - f(x_{k-1}, y)| \leq \int \sum_{k=1}^{n} |f(x_k, y, z) - f(x_{k-1}, y, z)|dz$$

$$\leq \int V(f(\cdot, y, z)) dz, \qquad (4.6)$$

where $V(f(\cdot, y, z))$ denotes the total variation of the function $x \to f(x, y, z)$. Taking the supremum over all possible $x_0 < \ldots < x_n$ in (4.6) and using Proposition 4.5.e) completes the proof of (4.4).

That equality holds in (4.5) if X and Y are independent follows directly from the definition of conditional variation. To see that independence is not necessary, consider four random variables X_1, X_2, Y_1 and Y_2 with continuously differentiable densities, such that X_1 is independent of Y_1, X_2 is independent of Y_2 and $EX_1 \neq EX_2$, $EY_1 \neq EY_2$. Further assume that the densities of X_1 and X_2 are unimodal, both having the same mode. For $0 \leq \lambda \leq 1$ define

$$(X_\lambda, Y_\lambda) = \begin{cases} (X_1, Y_1), & \text{with probability } \lambda, \\ (X_2, Y_2), & \text{with probability } (1 - \lambda). \end{cases}$$

A calculation from first principles using Proposition 4.5.d) shows that for all λ in $[0,1]$, $V(X_\lambda) = V_1(X_\lambda, Y_\lambda) = \lambda V(X_1) + (1 - \lambda)V(X_2)$. The covariance between X_λ and Y_λ is equal to $\lambda(1 - \lambda)E(X_1 - X_2)E(Y_1 - Y_2)$ so that X_λ and Y_λ are dependent (further, correlated) for $0 < \lambda < 1$.

The hypotheses of the previous paragraph are satisfied, in particular, when Y_1 and Y_2 have continuously differentiable densities and X_1 and X_2 follow Beta distributions with parameters $\alpha = 2$, $\beta = 3$ and $\alpha = 4$, $\beta = 7$, respectively. $\qquad \Box$

In Proposition 4.7 it is shown that the random vector $(tX)(\bmod 1)$ has a density if X has one.

Proposition 4.7 *Let X be an n dimensional random vector with density $f(x_1, \ldots, x_n)$. Then the function*

$$f_t(x_1, \ldots, x_n) = \frac{1}{t} \sum_{k \in \mathbb{Z}^n} f\left(\frac{x_1 + k_1}{t}, \ldots, \frac{x_n + k_n}{t}\right); \qquad 0 \leq x_1, \ldots, x_n \leq 1,$$

defines a density for X.

Proof. Similar to that of Proposition 3.4. $\qquad \Box$

Lemma 4.8 *If $X = (X_1, \ldots, X_n)'$ is an absolutely continuous random vector and U_n and U denote distributions uniform on $[0,1]^n$ and $[0,1]$ respectively, then:*

$$d_V\left(X(\text{mod}\,1),\, U_n\right) \le \mathbb{E}_X d_V\left((X_n|X_1, \ldots, X_{n-1})(\text{mod}\,1),\, U\right)$$

$$+ d_V\left((X_1, \ldots, X_{n-1})(\text{mod}\,1),\, U_{n-1}\right). \tag{4.7}$$

Proof. Propositions 2.5.b) and 4.7, and the definition of conditional density imply that

$$d_V\left(X(\text{mod}\,1),\, U_n\right) = \frac{1}{2}\int_{[0,1]^n}\left|\sum_{k \in \mathbf{Z}^n} f(x_n + k_n | x_1 + k_1, \ldots, x_{n-1} + k_{n-1})\right.$$

$$\left. \times f(x_1 + k_1, \ldots, x_{n-1} + k_{n-1}) - 1\right| dx_1 \ldots dx_n.$$

Adding and subtracting $\sum_{(k_1,\ldots,k_{n-1}) \in \mathbf{Z}^{n-1}} f(x_1 + k_1, \ldots, x_{n-1} + k_{n-1})$ within the absolute value and applying the triangular inequality leads to (4.7). □

The following is one of the main results of this chapter. It gives a tractable upper bound for the variation distance between $(tX)(\text{mod}\,1)$ and a distribution uniform on $[0,1]^n$.

Theorem 4.9 *Let $X = (X_1, \ldots, X_n)$ be an n dimensional absolutely continuous random vector such that every X_i has bounded mean-conditional variation with respect to the vector composed of the remaining X_j's. Define*

$$S(X) = \min_{\pi \in S_n} \sum_{i=1}^{n} V_1(X_{\pi(i)}, X_{\pi(1)}, \ldots, X_{\pi(i-1)}),$$

where the minimum is taken over all permutations π of the integers $\{1, \ldots, n\}$ and V_1 is defined in (4.3) above.

Then:

$$d_V\left((tX)(\text{mod}\,1),\, U_n\right) \le \frac{S(X)}{8t}. \tag{4.8}$$

Therefore $(tX)(\text{mod}\,1)$ converges to U_n at a rate at least linear in t^{-1}.

Proof. Due to Proposition 4.5.a) there is no loss of generality in assuming $t = 1$. Proceeding by induction on the dimension of the random vector X it is shown that

$$d_V\left(X(\text{mod}\,1),\, U_n\right) \le \frac{1}{8}\sum_{i=1}^{n} V_1(X_i, X_1, \ldots, X_{i-1}). \tag{4.9}$$

The one dimensional case follows from Theorem 3.9.b).

Assuming (4.9) holds for all $(n-1)$ dimensional random vectors, it is now shown that it holds for n dimensional distributions. Apply Lemma 4.8 and obtain upper bounds for the resulting right hand side using Theorem 3.9.b) for the first term and the induction hypothesis for the second term. This proves (4.9).

The fact that the roles of the X_i's can be permuted in (4.9) completes the proof.   □

Corollary. *Assume X and Y are n dimensional random vectors such that the conditional density of X given $Y = y$ satisfies the hypothesis of Theorem 4.9 for y in a set of probability one with respect to Y. Let*

$$S_1(X,Y) = \min_{\pi \in S_n} \sum_{i=1}^{n} E_Y V_1(X_{\pi(i)}, X_{\pi(1)}, \ldots, X_{\pi(i-1)} \mid Y).$$

Then:

$$d_V\big((tX + Y)(\text{mod}\,1),\, U_n\big) \leq \frac{S_1(X,Y)}{8t}.$$

Proof. Follows from Proposition 2.8.a) and Theorem 4.9.   □

Remark 1. The hypothesis that the mean-conditional variation of every random variable X_i with respect to the vector composed of the remaining X_j's is finite can be weakened by assuming that, eventually after reindexing the X_i's, the quantities $V(X_1), V_1(X_2, X_1), \ldots, V_1(X_n, X_1, \ldots, X_{n-1})$ are all finite. This ensures that at least one of the terms over which the minimum that defines $S(X)$ is taken is finite and (4.8) applies.

Remark 2. Proposition 4.6 implies that, given a random vector X with fixed marginals, the best possible bounds provided by (4.8) are

$$d_V\big((tX)(\text{mod}\,1),\, U_n\big) \leq \frac{1}{8t} \sum_{i=1}^{n} V(X_i), \tag{4.10}$$

where $V(X_i)$ denotes the (one dimensional) total variation of the random variable X_i.

Equation (4.10) holds, in particular, when the X_i's are independent. In this sense the bounds are smallest for independent random variables.

Remark 3. The bounds of Theorem 4.9 are sharp, that is, there exist random vectors for which they can be approached arbitrarily closely. Let $X_1, \ldots, X_n$ be independent random variables with X_1 uniform on $[0, k + \frac{1}{2}]$ and the remaining X_i's uniform on $[0, k^2 + \frac{1}{2}]$, k a positive integer.

Let $f(x_i)$ denote the density of $X_i(\text{mod}\,1)$, $i = 1, \ldots, n$. Then

$$f(x_1) = \begin{cases} (k+1)/(k + \frac{1}{2}), & 0 < x < \frac{1}{2}, \\ k/(k + \frac{1}{2}), & \frac{1}{2} < x < 1, \end{cases} \tag{4.11}$$

and for $i \geq 2$:

$$f(x_i) = \begin{cases} (k^2+1)/(k^2 + \frac{1}{2}), & 0 < x < \frac{1}{2}, \\ k^2/(k^2 + \frac{1}{2}), & \frac{1}{2} < x < 1. \end{cases} \tag{4.12}$$

Let $\mathbb{Z}_2^n$ denote the set of all n-tuples of zeros and ones and denote a typical element of this set by $\alpha = (\alpha_1 \ldots, \alpha_n)$. Proposition 2.5.b, (4.11) and (4.12) imply that

$$d_V\left(X(\mathrm{mod}\,1), U_n\right) = \frac{1}{2}\int |\Pi f(x_i) - 1|\,dx$$

$$= \frac{1}{2}\sum_{\alpha\in\mathbb{Z}_2^n}\int_{\Pi[\alpha_i/2,(\alpha_i+1)/2]} |\Pi f(x_i) - 1|\,dx$$

$$= \frac{1}{2^{n+1}}\sum_{\alpha\in\mathbb{Z}_2^n}\left|\frac{(k+\alpha_1)\Pi_{i\geq 2}(k+\alpha_i)}{(k+\frac{1}{2})(k^2+\frac{1}{2})^{n-1}} - 1\right|$$

$$= \frac{1}{2^{n+1}}\sum_{\alpha\in\mathbb{Z}_2^n}\left|\frac{(k+\alpha_1)\Pi_{i\geq 2}(k^2+\alpha_i) - (k+\frac{1}{2})(k^2+\frac{1}{2})^{n-1}}{(k+\frac{1}{2})(k^2+\frac{1}{2})^{n-1}}\right|.$$

The numerator of each of the 2^n terms in the sum is equal to $\frac{1}{2}k^{2n-2} + O(k^{2n-3})$ while the denominator is equal to $k^{2n-1} + O(k^{2n-2})$. Therefore the variation distance from $X(\mathrm{mod}\,1)$ to U_n is equal to $1/4k + O(1/k^2)$.

Theorem 4.9 gives the following upper bound:

$$\frac{1}{8}\sum_{i=1}^n V(X_i) = \frac{1}{8}\left\{\frac{2}{k+\frac{1}{2}} + \frac{2(n-1)}{k^2+\frac{1}{2}}\right\}$$

$$= \frac{1}{4k} + O\left(\frac{1}{k^2}\right).$$

It follows that the ratio between the bounds from Theorem 4.9 and the variation distance between $X(\mathrm{mod}\,1)$ and U_n can be arbitrarily close to one if k is sufficiently large.

Remark 4. Let b denote a fixed vector in $\mathrm{I\!R}^n$. Theorem 4.9 and Proposition 4.5.a) imply that

$$d_V\left((tX+b)(\mathrm{mod}\,1), U_n\right) \leq \frac{S(X)}{8t}.$$

Remark 5. Remark 2 following Theorem 3.9 and an argument similar to that used in the proof of Theorem 4.9 imply that $d_V\left((tX)(\mathrm{mod}\,1), U_n\right)$ tends to zero at least as fast as $t^{-(k+1)}$ if the density of $X = (X_1,\ldots,X_n)$, $f(x_1,\ldots,x_n)$, has k^{th} order partial derivatives with bounded variation. Hence, convergence is faster than any polynomial rate if $f(x_1,\ldots,x_n)$ is analytic and its partial derivatives of all orders are integrable.

Remark 6. Proposition 4.5.a) and Theorem 4.9 imply that for positive real numbers $t_1,\ldots,t_n$:

$$d_V\left(((t_1 X_1)(\mathrm{mod}\,1),\ldots,(t_n X_n)(\mathrm{mod}\,1)), U_n\right)$$

$$\leq \frac{1}{8}\min_{\pi\in S_n}\sum_{i=1}^n \frac{V_1(X_{\pi(i)}, X_{\pi(1)},\ldots,X_{\pi(i-1)})}{t_{\pi(i)}}.$$

Remark 7. From the fact that $x(\mathrm{mod}\,c) = c\{(x/c)(\mathrm{mod}\,1)\}$, Proposition 4.5.a) and Theorem 4.9 it follows that, for every vector $\tau = (\tau_1, \ldots, \tau_n)$ with positive components,

$$d_V\big((tX)(\mathrm{mod}\,\tau), U[0,\tau]\big) \leq \frac{1}{8t} \min_{\pi \in S_n} \sum_{i=1}^{n} \tau_{\pi(i)} V_1\big(X_{\pi(i)}, X_{\pi(1)} \ldots, X_{\pi(i-1)}\big),$$

where $U[0,\tau]$ denotes a distribution uniform on $\Pi_{i=1}^{n}[0,\tau_i]$ and $(\mathrm{mod}\,\tau)$ means that the ith component is considered $(\mathrm{mod}\,\tau_i)$.

Remark 8. It is not true that $(tX)(\mathrm{mod}\,1)$ converges to U_n faster when the expected values of the X_i's grow. The bounds in Theorem 4.9 are invariant under translation of X. It is not the magnitude of the values taken by the random vector X that matters, but its variability, as measured by its mean-conditional variation.   □

Even though the variation distance is a rather strong metric, when interested in the difference between the probability that $(tX)(\mathrm{mod}\,1)$ belongs to a certain set and the Lebesgue measure of that set, it does not take into account the set's size. A notion of distance which decreases with the size of the sets under consideration is the sup norm (see Proposition 2.10).

Bounds for the sup distance from $(tX)(\mathrm{mod}\,1)$ to a distribution uniform on $[0,1]^n$ are obtained in Theorem 4.12. They are four times those obtained for the variation distance plus higher order terms. In this sense, the higher dimensional result is analogous to the one dimensional one. Yet the assumptions under which they are derived are stronger.

The following definition and lemmas are used in the proof of Theorem 4.12.

Definition. Let X be a random vector with density $f(x_1, \ldots, x_n)$ and assume the crossed partial derivative $\frac{\partial^r f(x_1, \ldots, x_n)}{\partial x_{i_1} \ldots \partial x_{i_r}}$ is integrable in $\mathbb{R}^n$, with $1 \leq i_1 < \ldots < i_r \leq n$. The corresponding integral is denoted

$$V_{i_1, \ldots, i_r}(X) = \int_{\mathbb{R}^n} \left| \frac{\partial f(x_1, \ldots, x_n)}{\partial x_{i_1} \ldots \partial x_{i_r}} \right| dx_1 \ldots dx_n.$$

Lemma 4.10 *Let $g : \mathbb{R} \to \mathbb{R}$ be a continuous, piecewise differentiable, integrable, non negative function of bounded variation. Then*

$$\sum_{k \in \mathbb{Z}} g(x + k) \leq \int_{\mathbb{R}} g(x)dx + \frac{V(g)}{2},$$

where $V(g) = \int |g'(x)|dx$ denotes the total variation of $g(x)$.

Proof. Apply Theorem 3.9.c) to the random variable whose density evaluated at x is $g(x)/\int g(u)du$.   □

Lemma 4.11 *Assume the random vector $X = (X_1, \ldots, X_n)'$ has a density, $f(x)$, with continuous, piecewise differentiable, integrable, mixed partial derivatives of all orders* *less than or equal to n and that the set of zeros of these partial derivatives has no cluster point. Then*

$$\sum_{k \in \mathbb{Z}} f(x_n + k) V_{i_1,\ldots,i_r}(X_1,\ldots,X_{n-1}|X_n = x_n + k) \le V_{i_1,\ldots,i_r}(X_1,\ldots,X_n)$$

$$+ \frac{1}{2} V_{i_1,\ldots,i_r,n}(X_1,\ldots,X_n), \tag{4.13}$$

where the argument of $f(\cdot)$ indicates the density being considered and $1 \le i_1 < \ldots < i_r \le n - 1$.

Proof. Applying Lemma 4.10 to $x_n \to \left| \frac{\partial^r f(x_1,\ldots,x_n)}{\partial x_{i_1} \ldots \partial x_{i_r}} \right|$ (the hypothesis on the zeros of the partial derivative is needed to ensure this function is piecewise differentiable) and then integrating over $x_1,\ldots,x_n$ yields

$$\int \sum_k \left| \frac{\partial^r f(x_1,\ldots,x_{n-1},x_n+k)}{\partial x_{i_1} \ldots \partial x_{i_r}} \right| dx_1,\ldots,dx_{n-1}$$

$$\le \int \left| \frac{\partial^r f(x_1,\ldots,x_n)}{\partial x_{i_1} \ldots \partial x_{i_r}} \right| dx_1 \ldots dx_n + \frac{1}{2} \int \left| \frac{\partial^{r+1} f(x_1,\ldots,x_n)}{\partial x_{i_1} \ldots \partial x_{i_r} \partial x_n} \right| dx_1 \ldots dx_n.$$

As the right hand side of this inequality coincides with the right hand side of (4.13), all that remains to be shown is that the corresponding left hand sides are the same and this follows from the definition of conditional density, the fact that the i_p's are less than n and the Monotone Convergence Theorem. □

Theorem 4.12 *Assume the random vector X satisfies the hypothesis of Lemma 4.11 and denote*

$$S(X) = \min_{\pi \in S_n} \sum_{i=1}^n V_1(X_{\pi(i)}, X_{\pi(1)},\ldots,X_{\pi(i-1)}),$$

$$T_r(X) = \sum_{1 \le i_1 < \ldots < i_r \le n} V_{i_1,\ldots,i_r}(X); \qquad r = 2,\ldots,n.$$

Then:

$$\|(tX)(\bmod 1) - U_n\|_\infty \le \frac{S(X)}{2t} + \sum_{r=2}^n \frac{T_r(X)}{(2t)^r}. \tag{4.14}$$

Proof. It is easy to see that $V_{i_1,\ldots,i_r}(tX) = V_{i_1,\ldots,i_r}(X)/|t|^r$ and hence there is no loss of generality in assuming $t = 1$.

It is shown by induction on the dimension of the random vector X that

$$\|X(\bmod 1) - U_n\|_\infty \le \frac{S_1(X)}{2} + \sum_{r=2}^n \frac{T_r(X)}{2^r}, \tag{4.15}$$

where $S_1(X) = \sum_{k=1}^n V_1(X_k, X_1,\ldots,X_{k-1})$. Equation (4.14) then follows from the symmetric role played by the X_i's.

The one dimensional case follows from Theorem 3.9.c).

Assume (4.15) holds for all $(n-1)$ dimensional random vectors. Applying the triangular inequality:

$$\left| \sum_{k \in \mathbf{Z}^n} f(x_1 + k_1, \ldots, x_n + k_n) - 1 \right|$$

$$\leq \sum_{k_n} f(x_n + k_n) \left| \sum_{k_1, \ldots, k_{n-1}} f(x_1 + k_1, \ldots, x_{n-1} + k_{n-1} | x_n + k_n) - 1 \right|$$

$$+ \left| \sum_{k_n} f(x_n + k_n) - 1 \right|$$

$$= \sum_{k_n} f(x_n + k_n) \| (X_1, \ldots, X_{n-1} | X_n = x_n + k_n)(\mathrm{mod}\, 1) - U_{n-1} \|_\infty$$

$$+ \| X_n(\mathrm{mod}\, 1) - U \|_\infty.$$

Bound the first term using the induction hypothesis, Lemma 4.11 and Proposition 4.6, and the second term applying Theorem 3.9.c). This completes the proof.   □

Remark 1. Comments similar to numbers 3 through 7 following Theorem 4.9 can be made here too. For example, it is now shown that the bounds in (4.14) are sharp.

Let $X_1, \ldots, X_n$ be independent, identically distributed distributions uniform on $[0, k + \varepsilon]$, k a large integer and ε a small positive constant. A straightforward calculation shows that

$$\| X(\mathrm{mod}\, 1) - U_n \|_\infty = \left(\frac{k+1}{k+\varepsilon} \right)^n - 1.$$

The bounds given by Theorem 4.12 are equal to $n/(k + \varepsilon) + O(1/k^2)$ (where, strictly speaking, the densities of the X_i's should be smoothed at 0 and 1 so as to satisfy the hypothesis of the theorem). Taking k sufficiently large and ε sufficiently small makes the ratio of the bound and the sup distance arbitrarily close to one.

Remark 2. Comparing the bounds obtained for the variation distance in Theorem 4.9 with those derived for the sup distance in Theorem 4.12 it is interesting to note that the dominating term in the latter case is equal to four times the former one.

Remark 3. An alternative approach for constructing bounds for the sup norm is to mimic what was done for the variation distance. Instead of working with the mean-conditional variation the concept that naturally arises is:

$$V_\infty(X, Y) = \sup_{y \in [0,1]^p} \sum_{k \in \mathbf{Z}^p} V(X | Y = y + k) f(y + k). \qquad (4.16)$$

Results obtained for the variation distance hold for the sup distance, replacing V_1's by V_∞'s and $\frac{1}{2}$'s by $\frac{1}{8}$'s. The approach that leads to bounds in terms of the mean-conditional variation was chosen instead of that involving expressions like (4.16) because $V_1(X, Y)$ has a natural probabilistic interpretation while $V_\infty(X, Y)$ does not.

4.1.3 Exact Rates of Convergence

In Sect. 4.1.2 it was shown that, if all random vectors X having bounded mean-conditional variation are considered, the variation distance between $(tX)(\mathrm{mod}\,1)$ and U_n tends to zero at a rate linear in t^{-1} and this rate cannot be improved upon. If particular families are considered, convergence may take place at faster rates. Proposition 4.13 gives upper and lower bounds for $d_V\big((tX)(\mathrm{mod}\,1),\,U\big)$ in terms of the Fourier coefficients of X. Propositions 4.14 shows that $(tX)(\mathrm{mod}\,1)$ converges to U_n exponentially fast when X has a multivariate normal distribution. In Proposition 4.15 a family of random vectors is exhibited for which $(tX)(\mathrm{mod}\,1)$ is equal to U_n once t passes a certain threshold.

Proposition 4.13 *Assume* $X = (X_1,\ldots,X_n)'$ *is an absolutely continuous random vector with characteristic function* $\widehat{f}(\lambda)$. *Then*

$$\frac{1}{2}\sup_{m\in\mathbb{Z}_*^n}|\widehat{f}(2\pi t m)| \;\leq\; d_V\big((tX)(\mathrm{mod}\,1),\,U_n\big) \;\leq\; \frac{1}{2}\sum_{m\in\mathbb{Z}_*^n}|\widehat{f}(2\pi t m)|,$$

where $\mathbb{Z}_*^n$ *denotes the set of all n-tuples of integers with at least one component different from zero.*

Proof. The lower bound follows from Proposition 2.7 and Lemma 4.1.

The upper bound is a consequence of the n dimensional version of Poisson's Summation Formula (see Stein and Weiss, 1971, Corollary 1.8, p.249). □

Proposition 4.14 *Assume* $X = (X_1,\ldots,X_n)'$ *has a multivariate normal distribution. Then*

$$\frac{1}{2}\sup_{m\in\mathbb{Z}_*^n} e^{-2\pi^2 t^2 \mathrm{Var}(m\cdot X)} \;\leq\; d_V\big((tX)(\mathrm{mod}\,1),\,U_n\big)$$

$$\leq\; \min_{\tau\in S_n}\sum_{i=1}^{n} e^{-2\pi^2 t^2 \mathrm{Var}(X_{\tau(i)}|X_{\tau(1)},\ldots,X_{\tau(i-1)})} + \text{ higher order terms,} \quad (4.17)$$

where Var *denotes variance, $m\cdot X$ the usual inner product, S_n the set of permutations of $\{1,\ldots,n\}$ and $(X_{\tau(i)}|X_{\tau(1)},\ldots,X_{\tau(i-1)})$ the corresponding conditional distribution, which in the case of a multivariate normal is normal.*

Proof. The lower bound follows from Proposition 4.13. To obtain the upper bound, proceed as in the proof of Theorem 4.9, using Corollary 1 of Proposition 3.11 instead of Theorem 3.9.b). □

Remark 1. Given a bivariate normal distribution, assume the variances of $X, Y, (X|Y)$ and $(Y|X)$ are all different and denote the third largest of them by s^2. Equation (4.17) simplifies to

$$\frac{1}{2} \sup_{m \in \mathbb{Z}_*^2} e^{-2\pi^2 t^2 \mathrm{Var}(m_1 X + m_2 Y)} \leq d_V\big((tX)(\mathrm{mod}\,1),\, U_2\big)$$

$$\leq e^{-2\pi^2 t^2 s^2} + \text{higher order terms},$$

where $\mathbb{Z}_*^2$ denotes the set of ordered pairs of integers excluding $(0,0)$.

Remark 2. If the X_i's are independent, (4.17) becomes

$$\frac{1}{2} e^{-2\pi^2 t^2 s^2} \leq d_V\big((tX)(\mathrm{mod}\,1),\, U_n\big) \leq k e^{-2\pi^2 t^2 s^2} + \text{higher order terms},$$

where $s^2 = \min_{1 \leq i \leq n} \mathrm{Var}(X_i)$ and k denotes the number of X_i's attaining the minimum.

Remark 3. The variance of a conditional distribution is less than or equal to the variance of the original distribution. It follows that for random variables with fixed marginals the bounds in Proposition 4.14 are smallest when the random variables are independent. Remark 1 can be used to construct an example where the same rate of convergence achieved in the case of independence is attained in a situation where both random variables are highly correlated: choose $\mathrm{Var}(X) < \mathrm{Var}(Y|X)$ or $\mathrm{Var}(Y) < \mathrm{Var}(X|Y)$.

Remark 4. Define

$$s^2 = \max_{\pi \in S_n} \min_{1 \leq i \leq n} \mathrm{Var}(X_{\pi(i)} | X_{\pi(1)}, \ldots, X_{\pi(i-1)}). \tag{4.18}$$

The upper bound in (4.17) may be rewritten as

$$d_V\big((tX)(\mathrm{mod}\,1),\, U_n\big) \leq c e^{-2\pi^2 t^2 s^2} + \text{higher order terms},$$

where c is an integer that usually is equal to one and always is less than or equal to n.

Assume that, eventually after reindexing the X_i's, $s^2 = \mathrm{Var}(X_k | X_1, \ldots, X_{k-1})$. The upper and lower bounds in Proposition 4.14 have the same exponents (and therefore give the exact rate of convergence) if the coefficients of the regression of X_k on $X_1, \ldots, X_{k-1}$ are all integers. A particular case is when the X_i's are independent. Another case is when $s^2 = \mathrm{Var} X_i$, for some $i = 1, \ldots, n$.

Remark 5. The expression defined in (4.18) seems to have considerable structure, yet the author has not been able to give it a probabilistic interpretation (except in the bivariate case: see the first remark above).

Remark 6. All higher order terms mentioned above can be made explicit and the bounds are therefore not only asymptotic.

Proposition 4.15 *Let X be an n dimensional random vector. If the characteristic function of X has compact support, there exists t_0 such that $(tX)(\mathrm{mod}\,1)$ is equal to a distribution uniform on $[0,1]^n$ when $t \geq t_0$.*

Proof. Let $\widehat{f}(\lambda)$ denote the characteristic function of X. Lemma 4.1 and Proposition 2.1 imply that $(tX)(\mathrm{mod}\,1)$ is uniform for $t \geq t_0$ if and only if $\widehat{f}(2\pi tm) = 0$ for every $m \in \mathbb{Z}_*^n$ and every $t \geq t_0$. This is clearly the case if $\widehat{f}(\lambda)$ has compact support. $\qquad\square$

4.2 Applications

4.2.1 Lagrange's Top and Integrable Systems

Consider the motion of an axially symmetric body in a uniform gravitational field when one point of the symmetry axis is fixed in space. A variety of physical systems, ranging from a child's top to complicated gyroscopic navigational instruments, are approximated by such a *heavy symmetrical* or *Lagrangian* top.

The top's position at any instant in time can be described using three quantities called Euler angles (see Fig. 4.1). They are the rotation angle of the top about its own axis (ψ), the angle between the top's axis of symmetry and the vertical axis (θ) and the angle describing how the top moves around the vertical axis (ϕ). The top's movement is therefore composed of the superposition of three motions called rotation (ψ), nutation (θ) and precession (ϕ).

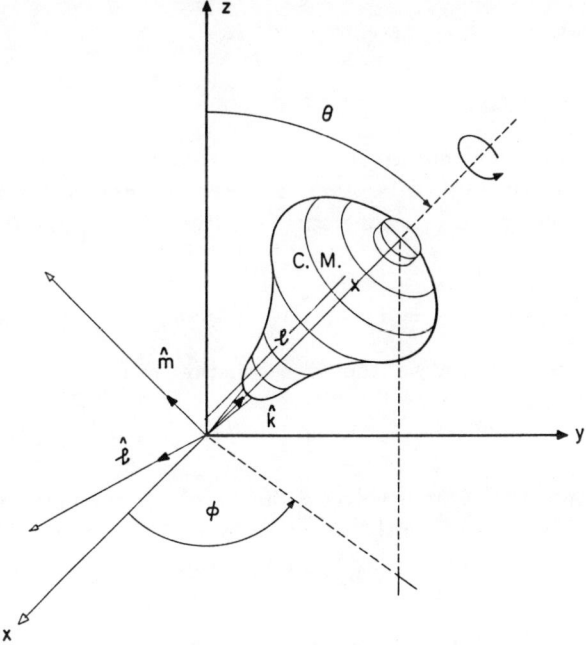

Fig. 4.1. Lagrangian top

The equations of motion for the top are now derived.

Let $\hat{k}$, $\hat{l}$, $\hat{m}$ denote a reference system that moves with the top (see Fig. 4.1). Since the top is symmetric, its kinetic energy may be written as

$$T = \frac{1}{2}I(\omega_1^2 + \omega_2^2) + \frac{1}{2}I_3\omega_3^2, \tag{4.19}$$

where I denotes the top's moment of inertia with respect to the $\hat{l}$ axis (which is equal to its moment of inertia with respect to the $\hat{m}$ axis), I_3 its moment of inertia with respect to the $\hat{k}$ axis, and ω_1, ω_2, ω_3 the components of its angular velocity along the $\hat{l}$, $\hat{m}$ and $\hat{k}$ axis, respectively.

Expressing (4.19) in terms of the Euler angles gives

$$T = \frac{1}{2}I(\dot{\theta}^2 + \dot{\phi}\sin^2\theta) + \frac{1}{2}I_3(\dot{\psi} + \dot{\phi}\cos\theta)^2,$$

where $\dot{x}$ denotes the time derivative of x.

As the top is moving in a gravitational field, its potential energy, V, may be calculated assuming all its mass is concentrated at its center of mass (C.M. in Fig. 4.1). This leads to

$$V = Mgl\cos\theta,$$

where M is the top's mass, g acceleration due to gravity and l the distance between the top's fixed point and its center of mass (see Fig. 4.1).

The top's Lagrangian, $\mathcal{L}$, is therefore equal to

$$\mathcal{L} = T - V = \frac{1}{2}I(\dot{\theta}^2 + \dot{\phi}\sin^2\theta) + \frac{1}{2}I_3(\dot{\psi} + \dot{\phi}\cos\theta)^2 - Mgl\cos\theta.$$

Because the Lagrangian does not involve ϕ or ψ explicitly, the corresponding Lagrange equations lead to two constants of motion. They correspond to the components of the angular momentum along the vertical z axis (see Fig. 4.1) and the $\hat{k}$ axis:

$$L_z = I\dot{\phi}\sin^2\theta + I_3\omega_3\cos\theta, \tag{4.20}$$

$$L_3 = I\omega_3 = I(\dot{\psi} + \dot{\phi}\cos\theta). \tag{4.21}$$

Because the system is conservative, the third constant of motion is its total energy:

$$E = T + V = \frac{1}{2}I(\dot{\theta}^2 + \dot{\phi}\sin^2\theta) + \frac{1}{2}I_3(\dot{\psi} + \dot{\phi}\cos\theta)^2 + Mgl\cos\theta. \tag{4.22}$$

A differential equation that only involves θ can be obtained using (4.20) and (4.22) to eliminate $\dot{\phi}$. This leads to a differential equation for $\dot{\theta}(t)^2$ in terms of a cubic polynomial involving $\cos\theta$. It is solved using elliptic integrals and the solutions for ψ and ϕ are obtained form (4.20) and (4.21).

For a rapidly spinning top, slow precession and nutation are frequently observed. Approximate solutions for the equations of motion are now derived for this case, following Barger and Olsson (1973).

Differentiating (4.20) with respect to time and neglecting the term involving $\dot{\phi}\dot{\theta}$ leads to

$$\ddot{\phi}\sin\theta = \frac{I_3\omega_3}{I}\dot{\theta}, \tag{4.23}$$

where $\ddot{x}$ denotes the second time derivative of x.

Differentiating (4.22) with respect to time, neglecting the term with $\dot{\phi}^2\dot{\theta}$ and using (4.23) gives

$$\ddot{\theta} = \left(\frac{Mgl}{I} - \frac{I_3\omega_3}{I}\dot{\phi}\right)\sin\theta. \tag{4.24}$$

Let

$$\omega_p = \frac{Mgl}{I_3\omega_3}, \qquad \omega_l = \frac{I_3\omega_3}{I}.$$

Differentiating (4.23) with respect to time, substituting in (4.24) and neglecting the term in $\ddot{\phi}\dot{\theta}$ gives

$$\frac{d^2\dot{\phi}}{dt^2} + \omega_l^2\dot{\phi} = \omega_l^2\omega_p. \tag{4.25}$$

Consider the following set of initial conditions:

$$\phi(0) = 0, \quad \dot{\phi}(0) = \omega_0, \quad \theta(0) = \theta_0, \quad \dot{\theta}(0) = 0, \quad \psi(0) = 0. \tag{4.26}$$

The only effect of assuming $\phi(0)$, $\psi(0)$, and $\dot{\theta}(0)$ equal to zero is that it simplifies some of the constants that follow. The conclusions that are derived remain valid when these quantities have arbitrary initial values.

Let

$$A = \frac{\omega_p - \omega_0}{\omega_l}.$$

Equations (4.23) and (4.26) imply that $\ddot{\phi}(0) = 0$. Using this when solving (4.25) yields

$$\dot{\phi}(t) = \omega_p + (\omega_0 - \omega_p)\cos(\omega_l t), \tag{4.27}$$

and hence

$$\phi(t) = \omega_p t - A\sin(\omega_l t). \tag{4.28}$$

Substituting (4.28) into (4.23) and approximating $\sin\theta$ by $\sin\theta_0$ – nutation has been assumed to vary slowly – leads to

$$\theta(t) = \theta_0 + A\sin\theta_0 - A\cos(\omega_l t). \tag{4.29}$$

Finally, substituting (4.17) in (4.21) and taking $\cos\theta \simeq \cos\theta_0$ gives:

$$\psi(t) = \omega_r t + A\cos\theta_0\sin(\omega_l t), \tag{4.30}$$

where

$$\omega_r = \omega_3 - \omega_p\cos\theta_0.$$

Equations (4.28), (4.29) and (4.30) provide approximate solutions for the equations of motion of a Lagrangian top when precession and nutation vary slowly. They show that the top's position at time t, as described by $(\psi(t)(\mathrm{mod}\,2\pi), \theta(t), \phi(t)(\mathrm{mod}\,2\pi))$, is a function of $\omega_p t(\mathrm{mod}\,2\pi)$, $\omega_l t(\mathrm{mod}\,2\pi)$, $\omega_r t(\mathrm{mod}\,2\pi)$, ω_p, $(\omega_p - \omega_0)/\omega_l$ and θ_0. If the physical constants that appear in the solution of the equations of motion are not known exactly and described by a joint density, Theorems 4.2 and 4.4 and the Continuous Mapping Theorem imply that the random vector describing the top's Euler angles at time t has the following limit (in the weak-star topology, as t tends to infinity):

$$\begin{pmatrix} \phi(t)(\mathrm{mod}\, 2\pi) \\ \theta(t) \\ \psi(t)(\mathrm{mod}\, 2\pi) \end{pmatrix} \rightarrow \begin{pmatrix} (U_1 - A\sin U_2)(\mathrm{mod}\, 2\pi) \\ (\theta_0 + A\sin\theta_0 - A\cos U_2)(\mathrm{mod}\, 2\pi) \\ (U_3 + A\cos\theta_0 \sin U_2)(\mathrm{mod}\, 2\pi) \end{pmatrix},$$

where U_1, U_2 and U_3 are independent, identically distributed random variables, uniform on $[0, 2\pi]$. A calculation based on Proposition 2.1 and Lemma 4.1 shows that the limiting random vector has the same distribution as $(W_1, \theta_0 + A\sin\theta_0 - A\cos W_2, W_3)'$ where W_1, W_2, and W_3 are independent, identically distributed random variables uniform on $[0, 2\pi]$, independent of (A, θ_0).

Assume the top is painted half black and half white and that a circle is drawn on the surface where it is spinning, centered at its fixed point (see Fig. 4.2).

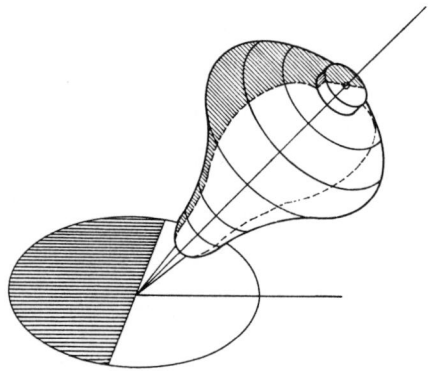

Fig. 4.2. Painted top

The top is allowed to spin long enough and stopped. The probability that the half of the top that faces the floor when it stops is mainly black will be approximately equal to one half. Similarly, the probability that the top's main axis is above the white half of the circle when it stops is also approximately one half. Further, the probability that both events just described take place simultaneously is near one forth. To obtain a rough estimate of how much time must pass before the probability of both events happening simultaneously is between 0.24 and 0.26, note that conditioning on the values of A, ω_l and θ_0, applying Theorem 4.9 and using Proposition 2.8.b) to uncondition leads to

$$\mathrm{d_V}\big((\phi(t), \psi(t))(\mathrm{mod}\, 2\pi), U\big) \le \frac{\pi \mathrm{ES}(\omega_p, \omega_r | \omega_l, A, \omega_0)}{4t}, \qquad (4.31)$$

where U denotes a distribution uniform on $[0, 2\pi]^2$, $\mathrm{S}(X)$ a measure of smoothness of the joint density of X (see Theorem 4.9) and E expectation with respect to (ω_l, A, ω_0).

Typical values of $\omega_p = Mgl/I_3\omega_3$ and $\omega_r = \omega_3 - \omega_p \cos\theta_0$ are needed to find an upper bound for $\mathrm{ES}(\omega_p, \omega_r | \omega_l, A, \omega_0)$. Assume the top's mass is one pound and the distance from its fixed point to its center of mass, l, is six inches. Its moment of inertia with respect to its axis of symmetry, I_3, may be approximated by that of a solid sphere of radius l and therefore $I_3 \simeq 0.1$ lb·ft^2. The top is rotating fast, so a typical value of its initial angular velocity is, say, 6π rad/sec. The angle it initially makes with the vertical axis is, say, $\pi/12$ radians. Then $\omega_p \simeq 8.5$ rad/sec and $\omega_r \simeq 11.5$ rad/sec. A conservative upper bound for $S(\omega_p, \omega_r | \omega_l, A, \omega_0)$ is therefore equal to three times the corresponding expression for a distribution uniform on $[6, 12] \times [8, 14]$ that does not depend on (ω_l, A, ω_0). This bound and (4.31) lead to

$$d_V\big((\phi(t), \psi(t))(\mathrm{mod}\, 2\pi),\, U\big) \;\leq\; \frac{\pi}{2t}.$$

Therefore the probability that the half of the top that faces the surface is mainly black and that its axis of symmetry is above the black half of the circle is between 0.24 and 0.26 after 158 seconds. The same probability belongs to $[0.20,\, 0.30]$ after 32 seconds.

Bounds similar to those obtained in (4.31) can be derived for the variation distance between the three Euler angles needed to describe the top's position and the limiting distribution discussed above.

It is interesting to note that the angle $\theta(t)$ describing the top's nutation does not approach a uniform distribution as time passes. Conditional on the value of A and θ_0 its limiting distribution is an arcsine law. There is a larger probability of finding $\theta(t)$ near its maximum and minimum values than finding it near intermediate ones.

The Lagrangian top is an example of an integrable system, that is, a physical system with Hamiltonian equations solvable by the method of quadratures. Denote by $\xi_1(t), \ldots, \xi_n(t)$ the variables describing the position of an arbitrary integrable system and assume they take values on a compact set. Liouville's theorem for integrable systems (see Arnold, 1978, p.271ff) implies that there exists a smooth (diffeomorphic) change of variables $\xi = (\xi_1, \ldots, \xi_n) \to \eta = (\eta_1, \ldots, \eta_n)$ such that

$$\eta_i(t)(\mathrm{mod}\, 1) = \big(\eta_i(0) + \dot{\eta}_i(0)t\big)(\mathrm{mod}\, 1)\,; \qquad i = 1, \ldots, n. \tag{4.32}$$

In the case of the approximate analysis carried out above for the Lagrangian top, $n = 3$, $\xi_1 = \phi$, $\xi_2 = \theta$ and $\xi_3 = \psi$, and the η's are given by

$$\eta_1(t) = \phi(t) + A\sin(\omega_l t),$$

$$\eta_2(t) = \cos^{-1}\left(\frac{\theta_0 + A\sin\theta_0 - \theta(t)}{A}\right),$$

$$\eta_3(t) = \psi(t) - A\cos\theta_0 \sin(\omega_l t).$$

Equations (4.28), (4.29) and (4.30) imply that

$$\eta_1(t) = \omega_p t, \qquad \eta_2(t) = \omega_l t, \qquad \eta_3(t) = \omega_r t,$$

where $\omega_p = Mgl/I_3\omega_3$, $\omega_l = I_3\omega_3/I$ and $\omega_r = \omega_3 - \omega_p \cos\theta_0$ (see the beginning of this section for the notation used above). The quantities ω_p, ω_l and ω_r are called precession, nutation and rotation frequencies, respectively.

Liouville's theorem for integrable systems does not have a constructive proof and therefore the change of variables leading to (4.32) is usually not known explicitly.

Assume initial conditions are described using a joint density and none of the ξ_i's is redundant (non degeneracy is the precise concept). Then $\dot{\eta}(0) = (\dot{\eta}_1(0), \ldots, \dot{\eta}_n(0))$ also has a density and Theorem 5.3 implies that $\eta(t)(\text{mod } 1)$ converges in the variation distance to a distribution uniform on $[0,1]^n$. The new variables describing the system become independent as time passes. The change of variables leading to the η's is invertible and hence (Proposition 2.6) the vector describing the system's position originally also has a limit that does not depend on initial conditions. As the change of variable is usually not known explicitly, the distribution of this limit is not known either. Its coordinates are not necessarily independent. The (approximate) analysis carried out for the Lagrangian top is an exception: the change of variables is known and the limiting random variables are independent.

Though they are more the exception than the rule, there are many well known integrable systems. For example, the planar motion of a mass point under the gravitational attraction of two fixed mass points, a problem going back to Euler, is an integrable system. For other integrable systems, see Cornfeld, Fomin and Sinai (1982) and Moser (1973). Arnold (1978, p.273) points out that Liouville's theorem for integrable systems covers all problems of dynamics which have been integrated to the present day.

4.2.2 Coupled Harmonic Oscillators

A simple pendulum consists of a small weight, suspended by a light inextensible cord. When pulled to one side of its equilibrium position and released, the pendulum swings in a vertical plane under the influence of gravity.

Consider a pair of coupled harmonic oscillators, that is, two simple pendulums whose bobs are connected by a spring, as indicated in Fig. 4.3. For simplicity assume both bobs have the same mass, m, and both cords the same length, l. Let k' denote the spring's constant and $x_1(t)$ and $x_2(t)$ the weights' displacement from their equilibrium position at time t.

If initial displacements are small, the oscillator's kinetic and potential energy are approximately equal to

$$T = \frac{1}{2}m\big(\dot{x}_1^2(t) + \dot{x}_2^2(t)\big),$$

$$V = \frac{1}{2}k\big(x_1^2(t) + x_2^2(t)\big) + \frac{1}{2}k'\big(x_1(t) - x_2(t)\big)^2,$$

where $\dot{x}$ indicates the time derivative of x and $k = mg/l$.

Lagrange's equations imply that

$$m\ddot{x}_1(t) = -kx_1(t) - k'\big(x_1(t) - x_2(t)\big),$$
$$m\ddot{x}_2(t) = -kx_2(t) + k'\big(x_1(t) - x_2(t)\big).$$

The solutions to these equations are:

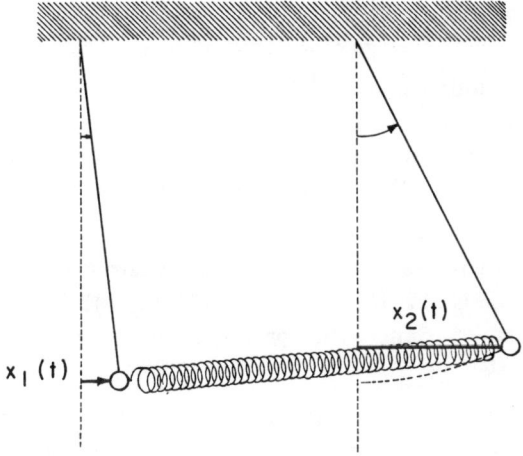

Fig. 4.3. Coupled harmonic oscillators

$$x_1(t) = A\cos(\omega t + \alpha) + B\cos(\omega' t + \beta),$$
$$x_2(t) = A\cos(\omega t + \alpha) - B\cos(\omega' t + \beta),$$

where $\omega = \sqrt{k/m}$ and $\omega' = \sqrt{(k + 2k')/m}$ and the constants A, B, α and β depend on the bobs' initial positions and velocities. For example, if both begin from rest $(\dot{x}_1(0) = \dot{x}_2(0) = 0)$:

$$x_1(t) = A\cos(\omega t) + B\cos(\omega' t), \quad x_2(t) = A\cos(\omega t) - B\cos(\omega' t), \tag{4.33}$$

with

$$A = \frac{x_1(0) + x_2(0)}{2}, \qquad B = \frac{x_1(0) - x_2(0)}{2}. \tag{4.34}$$

Now assume the values of the physical constants k, k' and m and those of the initial displacements are not known exactly but described by a five dimensional random vector.

From (4.33) and (4.34) (or from plain common sense) it is clear that the set of possible values of $x_1(t)$ and $x_2(t)$ depends on how large the initial displacements are. Therefore the dependence of $x_1(t)$ and $x_2(t)$ on their initial values cannot wash away as time passes. Dependence on the physical constants k, k' and m does wash away, as is now shown.

From a physical point of view it is natural to assume that (k, k', m) and $\big(x_1(0), x_2(0)\big)$ are independent, for there is no relation between, say, a bob's mass and its initial displacement. Note that the argument that follows can be easily adapted to the case of dependence.

Equation (4.33) and the Continuous Mapping Theorem imply that the random vector $\big(x_1(t), x_2(t)\big)$ converges in the weak-star topology to a random vector $L = (L_1, L_2)$ with $L_1 = AS_1 + BS_2$, $L_2 = AS_1 - BS_2$; S_1 and S_2 independent, identically distributed arcsine laws with common density

$$f(s) = \frac{1}{\pi\sqrt{1 - s^2}} \; ; \qquad\qquad -1 < s < 1.$$

Further, S_1 and S_2 are independent of A and B (Theorem 4.4). As time passes, the position of the coupled harmonic oscillator converges to a sum of mixtures of independent arcsine laws. The mixing random variables are determined by those describing the initial displacement of the bobs. Dependence on the physical constants k, k' and m washes away.

The limit random variables L_1 and L_2 both have zero mean and variance $\frac{1}{4}E\big(x_1^2(0) + x_2^2(0)\big)$. Their correlation is $2Ex_1(0)x_2(0)/(E\big(x_1^2(0) + x_2^2(0)\big)$. If $x_1(0)$ and $x_2(0)$ both have zero mean and the same variance, the correlation between L_1 and L_2 is equal to that between $x_1(0)$ and $x_2(0)$.

To obtain upper bounds on the variation distance between $\big(x_1(t), x_2(t)\big)$ and L, assume that the mean-conditional variation of ω with respect to ω' is finite (see Sect. 4.1.2 for the corresponding definition) and viceversa. This is the case, for example, if the partial derivatives of the joint density of ω and ω' are integrable. Then

$$d_V\left(\big((\omega t)(\mathrm{mod}\, 2\pi), (\omega' t)(\mathrm{mod}\, 2\pi)\big), U_2\right) \leq \frac{\pi c}{4t}, \qquad (4.35)$$

where $c = \min(V_1(\omega) + V_1(\omega', \omega), V_1(\omega') + V_1(\omega, \omega'))$ is a measure of how smooth the joint density of ω and ω' is (see Theorem 4.9).

The generalization of Theorem 5.2 discussed in Sect. 5.7 can be used to show that $\big(x_1(t), x_2(t)\big)$ converges in the variation distance to L for any density describing initial conditions. Bounds like those derived in (4.35) require additional smoothness assumptions.

A system that consists of n coupled harmonic oscillators is now considered (see Fig. 4.4).

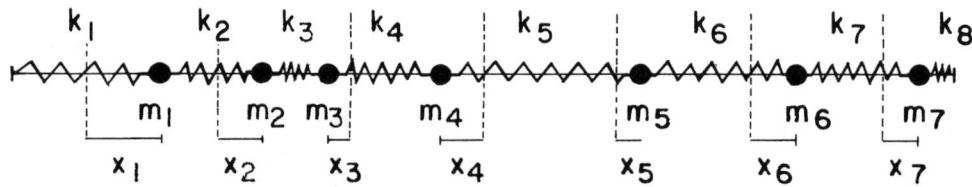

Fig. 4.4. System consisting of n coupled oscillators

Let $x_i(t)$ denote the displacement of the i-th bob from its equilibrium position at time t and k_i the spring constant of the i-th spring, $i = 1, \ldots, n + 1$ (see Fig. 4.4).

The system's kinetic energy is

$$T = \frac{1}{2}\sum_{i=1}^{n} m_i \dot{x}_i^2,$$

and its potential energy is

$$V = \frac{1}{2} \sum_{i=1}^{n+1} k_i (x_i - x_{i-1})^2, \tag{4.36}$$

where $x_0 = x_{n+1} = 0$.

Define the matrix

$$A = \begin{pmatrix} k_1 + k_2 & -k_2 & 0 & \cdots & 0 & 0 \\ -k_2 & k_2 + k_3 & -k_3 & \cdots & 0 & 0 \\ 0 & -k_3 & k_3 + k_4 & \cdots & 0 & 0 \\ \vdots & \vdots & \vdots & \ddots & \vdots & \vdots \\ 0 & 0 & 0 & \cdots & k_{n-1} + k_n & -k_n \\ 0 & 0 & 0 & \cdots & -k_n & k_n + k_{n+1} \end{pmatrix}.$$

and let R denote a diagonal matrix with $1/\sqrt{m_i}$ at the i-th position on its diagonal. Consider the "mass-weighted coordinates" $z_i = \sqrt{m_i} x_i$, $i = 1, \ldots, n$. In this new coordinate system the kinetic and potential energy are equal to

$$T = \frac{1}{2} \sum \dot{z}_i^2, \qquad V = \frac{1}{2} z' M z,$$

respectively, where $z = (z_1, \ldots, z_n)'$ and $M = RAR$.

Lagrange's equations are:
$$\ddot{z} = Mz. \tag{4.37}$$

Equation (4.36) implies that A is positive definite ($x'Ax > 0$ for $x \neq 0$). Since R is not singular, this implies that M is positive definite. Therefore there exists a diagonal matrix Λ with positive real numbers on its diagonal $(\lambda_1, \ldots, \lambda_n)$ (corresponding to the eigenvalues of M) and an orthogonal matrix P such that

$$M = P' \Lambda P,$$

where P' denotes the transpose of P.

Let $y = Pz = PR^{-1}x$. Equation (4.37) leads to

$$\ddot{y} = \Lambda y.$$

Hence:

$$y_i(t) = A_i \cos(\omega_i t + \phi_i), \qquad\qquad i = 1, \ldots, n,$$

where $\omega_i = \sqrt{\lambda_i}$ and $A_i = \sqrt{y_i^2(0) + (\dot{y}_i^2(0)/\lambda_i)}$.

Let

$$S(t) = \big(\cos(\omega_1 t + \phi_1), \ldots, \cos(\omega_n t + \phi_n) \big)',$$

and denote by A a diagonal matrix with i-th element on its diagonal equal to A_i. Then $x(t)$ may be written as

$$x(t) = RP' A S(t). \tag{4.38}$$

The spring constants and the bobs' masses are usually unknown and may be described by a joint density. Though the change of variables that leads to the angular frequencies $\lambda_1, \ldots, \lambda_n$ is quite complicated, physical considerations show that $\lambda_1, \ldots, \lambda_n$ are not restricted to a lower dimensional space, and therefore the n dimensional random vector $(\lambda_1, \ldots, \lambda_n)$ also has a density. Theorem 4.2 implies that $S(t)$ converges in the weak-star topology to a random vector whose components are independent, identically distributed arcsine laws with densities

$$f(s) = \frac{1}{\pi\sqrt{1-s^2}}, \qquad 0 < s < 1.$$

Theorem 4.4, (4.38) and the Continuous Mapping Theorem imply that the random vector of displacements, $x(t)$, converges in the weak-star topology to the product of a stochastic matrix $B = RP'A$ and a random vector of independent, identically distributed arcsine laws independent of B. A calculation from first principles shows that the expectation and covariance matrix of the limiting random vector are equal to the null vector and the matrix $\frac{1}{2}EBB'$, respectively.

As time passes, the random vector $x(t)$ approaches a limit that is partly determined by initial conditions and partly independent of them. This behavior is a particular case of "partial statistical regularity," a concept introduced and discussed in Sect. 5.7.

The motion of any conservative system in the neighborhood of a configuration of stable equilibrium exhibits a behavior of the type just described and may be analyzed in a similar way (see Goldstein, 1980, Chap. 6). It is known as the problem of "small oscillations." The only caveat is that in the general case the matrix M might have eigenvalues equal to zero, which from a physical point of view means that the system has certain symmetries and therefore has less than n degrees of freedom. In this case the conclusion derived for the coupled harmonic oscillator applies to a subset of the original variables: there exists a subset of k – out of the original n – variables that describes the system's position and such that the corresponding k dimensional random vector converges (in the weak-star topology) to the product of stochastic matrix $\widetilde{M}$ and a vector of independent, identically distributed arcsine laws independent of $\widetilde{M}$.

4.2.3 Billiards

In his classic "Théorie des Probabilités," published in 1909, Émile Borel dedicates a full chapter to the method of arbitrary functions (see Borel, 1965, for an English translation). Among the examples he considers is that of a point mass restricted to uniform rectilinear motion on the unit square and reflected in accordance with the laws of classical mechanics (see Fig. 4.5).

Borel shows that the random variable describing the particle's position approaches a distribution uniform on the unit square, if uncertainty about the particle's initial velocity is described by a uniform distribution on a rectangle.

In this section, Borel's results are extended to any joint density describing the particle's initial position and velocity. If this joint density is sufficiently smooth, upper bounds for the rate of convergence are provided.

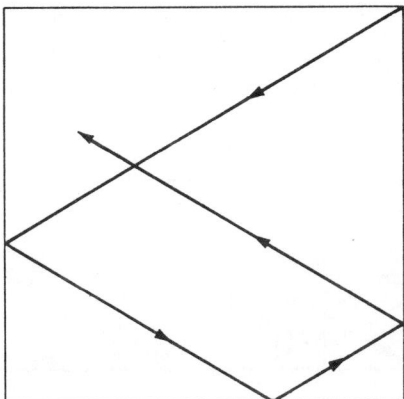

Fig. 4.5. Trajectory of particle

Driven by unrelenting scientific zeal, the author spent a full afternoon at a pool table located nearby Stanford University conducting experiments to provide a real life example. The corresponding results are presented at the end of this section.

Consider a billiard ball subject to uniform rectilinear motion, bouncing off the sides of the billiard table according to the laws of classical mechanics (angle of incidence = angle of reflection). That is, it is assumed that there is no friction slowing down the ball and collisions are perfectly elastic.

Without loss of generality assume the ball is released at time $t = 0$ from the point $(0,0)$. Denote its initial velocity by $v = (v_1, v_2)$ and its position at time t by $x(t) = (x_1(t), x_2(t))$.

To apply the theory developed in this chapter, an expression for the billiard ball's position at a given instant of time t, $x(t)$, in terms of $(tX)(\mathrm{mod}\,1)$ (where X is some random vector) must be found.

Define the auxiliary trajectory $y(t) = (y_1(t), y_2(t))$ as follows:

$$y_1(t) = v_1 t \quad ; \qquad y_2(t) = v_2 t\,.$$

Figure 4.6 is helpful to conclude that, at any instant of time, $x_i(t)$ is equal to either $y_i(t)(\mathrm{mod}\,1)$ or $1 - y_i(t)(\mathrm{mod}\,1)$, $i = 1, 2$, and therefore $x(t)$ is a function of $(vt)(\mathrm{mod}\,1)$. This observation is not enough to ensure that $x(t)$ converges to a distribution uniform on the unit square when t tends to infinity. Even though $\big((v_1 t)(\mathrm{mod}\,1),\, (v_2 t)(\mathrm{mod}\,1)\big)$, $\big((v_1 t)(\mathrm{mod}\,1),\, 1 - (v_2 t)(\mathrm{mod}\,1)\big)$, $\big(1 - (v_1 t)(\mathrm{mod}\,1),\, (v_2 t)(\mathrm{mod}\,1)\big)$ and $\big(1 - (v_1 t)(\mathrm{mod}\,1),\, 1 - (v_2 t)(\mathrm{mod}\,1)\big)$ all converge to a distribution uniform on the unit square, and at any instant in time $x(t)$ is equal to one of the four preceding random vectors, there could be an adverse selection process involved in determining which random vector is equal to $x(t)$ and the limit might not be uniform. It is therefore important to determine at which instants of time $x(t)$ is equal to any given random vector among those just mentioned.

Figure 4.6 is useful to see that $x_1(t)$ is equal to $(v_1 t)(\mathrm{mod}\,1)$ if and only if $(v_1 t)(\mathrm{mod}\,2) \le 1$. This observation, its counterpart for $x_2(t)$, and the fact that $r(\mathrm{mod}\,1)$

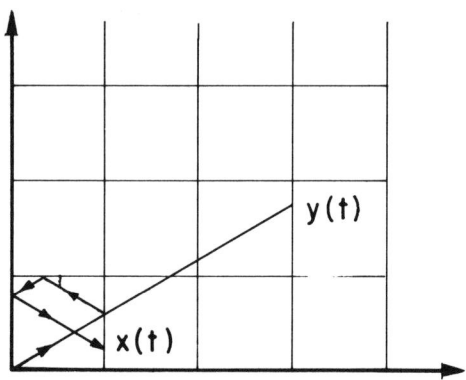

Fig. 4.6. Original and auxiliary trajectories

$= (r(\mathrm{mod}\,2))(\mathrm{mod}\,1)$, can be used to provide an explicit expression for $x(t)$ in terms of $(v_1 t)(\mathrm{mod}\,2)$ and $(v_2 t)(\mathrm{mod}\,2)$. Consider the following sets:

$$A_1 = \{t : (v_1 t)(\mathrm{mod}\,2) \le 1, (v_2 t)(\mathrm{mod}\,2) \le 1\},$$
$$A_2 = \{t : (v_1 t)(\mathrm{mod}\,2) \le 1, (v_2 t)(\mathrm{mod}\,2) > 1\},$$
$$A_3 = \{t : (v_1 t)(\mathrm{mod}\,2) > 1, (v_2 t)(\mathrm{mod}\,2) \le 1\},$$
$$A_4 = \{t : (v_1 t)(\mathrm{mod}\,2) > 1, (v_2 t)(\mathrm{mod}\,2) > 1\}.$$

Then:

$$(x_1(t), x_2(t)) = \begin{cases} \big((v_1 t)(\mathrm{mod}\,2), (v_2 t)(\mathrm{mod}\,2)\big)(\mathrm{mod}\,1), & \text{if } t \in A_1, \\ \big((v_1 t)(\mathrm{mod}\,2), 2 - (v_2 t)(\mathrm{mod}\,2)\big)(\mathrm{mod}\,1), & \text{if } t \in A_2, \\ \big(2 - (v_1 t)(\mathrm{mod}\,2), (v_2 t)(\mathrm{mod}\,2)\big)(\mathrm{mod}\,1), & \text{if } t \in A_3, \\ \big(2 - (v_1 t)(\mathrm{mod}\,2), 2 - (v_2 t)(\mathrm{mod}\,2)\big)(\mathrm{mod}\,1), & \text{if } t \in A_4. \end{cases} \quad (4.39)$$

Assume the characteristic function of the random vector (v_1, v_2) vanishes at infinity (e.g. that it has a density). Then $((v_1 t)(\mathrm{mod}\,2), (v_2 t)(\mathrm{mod}\,2))$ converges (in the weak-star topology) to a distribution uniform on $[0,2]^2$ (Theorem 4.2). Equation (4.39) combined with the Continuous Mapping Theorem (see e.g. Billingsley, 1986, p.391) imply that $x(t)$ converges to the function of a distribution uniform on $[0,2]^2$ implicitly defined in (4.39). A calculation from first principles shows that, as expected, this limit is a distribution uniform on the unit square. Further, Theorem 5.3 can be used to show that, for any density describing initial conditions, $x(t)$ converges to a uniform distribution in the variation distance.

Assume the random variable v_1 has bounded mean-conditional variation with respect to v_2 and viceversa (see Sect. 4.1.2 for the definition of mean-conditional variation). This is the case, for example, if the joint density of v_1 and v_2 has integrable partial derivatives (see Proposition 4.5.d)). Theorem 4.9 and Proposition 4.6 imply that

$$d_V\big(x(t), U_2\big) \le d_V\big(((v_1 t)(\mathrm{mod}\,2), (v_2 t)(\mathrm{mod}\,2)), \tilde{U}_2\big)$$

$$\leq \frac{\min\big(V(v_1) + V_1(v_2, v_1), V(v_2) + V_1(v_1, v_2)\big)}{4t}$$

$$\leq \frac{V_1(v_1, v_2) + V_1(v_2, v_1)}{4t}, \tag{4.40}$$

where U_2 and $\widetilde{U}_2$ denote distributions uniform on $[0,1]^2$ and $[0,2]^2$, respectively.

Suppose now that there also is some uncertainty about the ball's initial position $x(0) = (x_1(0), x_2(0))$. The Corollary of Theorem 4.9 gives bounds for the variation distance between $x(t)$ and a distribution uniform on the unit interval. In particular, if the initial position is independent of the velocity vector, which is a reasonable assumption, the bounds provided by (4.40) remain valid.

To apply this result to a real life situation, consider a standard billiard table. Its length and width are 9 and $4\frac{1}{2}$ feet, respectively. Let $\widetilde{U}$ denote a distribution uniform on $[0,9] \times [0, 4\frac{1}{2}]$. Equation (4.40) and basic properties of the variation distance imply that

$$d_V(x(t), \widetilde{U}) \leq \frac{2.25V_1(v_1, v_2) + 1.125V_1(v_2, v_1)}{t}. \tag{4.41}$$

Equation (4.41) and the fact that the total variation of a random variable is invariant under translation show that the bounds in (4.41) do not depend on how large the initial velocity is. However, they do depend on the variability of the random vector describing it. Experiments showed that typical initial velocities are in the range of 4 to 10 feet per second. Assume the ball is rolled making an initial angle of approximately 45 degrees with the table's longer side (see Fig. 4.7).

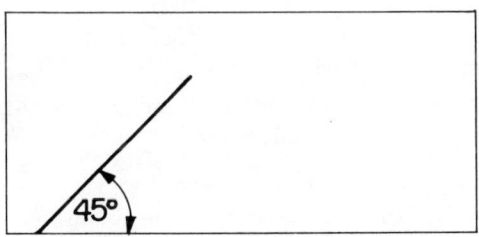

Fig. 4.7. Billiard table showing 45^0 angle

Both components of the velocity take values in a range between 3 and 7 feet/sec. The simplest assumption is to take v_1 and v_2 uniform in this range. Equation (4.41) implies that the variation distance between $x(t)$ and $\widetilde{U}$ is less than 0.05 after 33.8 seconds.

Next consider a simple model which allows for correlation between v_1 and v_2. Assume (v_1, v_2) has a distribution uniform on an ellipse centered at $(5, 5)$ with principal axis' of lengths a and b. As the ball is being rolled at an angle of approximately 45^0 with the table's longer side (x axis), one of the ellipse's axis, say that of length a, is assumed to make a 45^0 angle with the x axis (see Fig. 4.8).

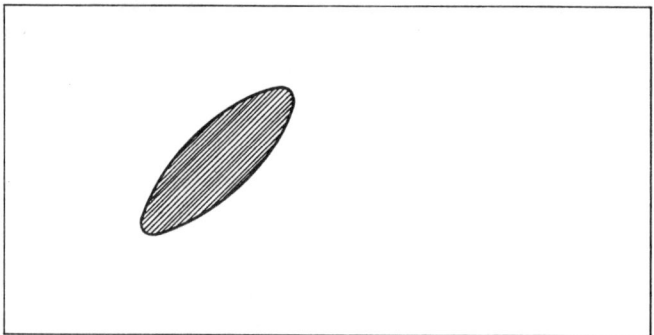

Fig. 4.8. Rotated ellipse

Example 2 following Proposition 4.5 shows that $V_1(v_1, v_2)$ and $V_1(v_2, v_1)$ are both equal to $2\sqrt{2(a^2 + b^2)}/\pi ab$. The correlation between both random variables is $(a^2 - b^2)/(a^2 + b^2)$. Since the person rolling the ball is attempting to let it go at a 45^0 degree angle, both initial velocities can be expected to be positively correlated. It is therefore assumed that the initial velocity is a uniform random variable on the rotated ellipse with $a = 4\frac{1}{2}$ and $b = 1\frac{1}{2}$. The correlation between v_1 and v_2 is then equal to 0.8 and the variation distance between $x(t)$ and $\tilde{U}$ is less than 0.05 after 42.7 seconds.

The approach used to obtain concrete bounds in the previous paragraphs can be extended to mixtures of unimodal random variables using Proposition 4.5.c).

Another interesting case is when the initial velocities v_1 and v_2 arise from a mixture of bivariate normal distributions where the variances of v_1 and v_2 are larger than 3 and 1 (ft/sec)^2, respectively, and their correlation coefficient, r, has absolute value less than 0.8. Remark 1 following Proposition 4.14 and the fact that $\mathrm{Var}(v_1|v_2) = (1 - r^2)\mathrm{Var}(v_1)$ imply that the variation distance between $x(t)$ and $\tilde{U}$ is less than or equal to $e^{-0.08\pi^2 t^2}$ and hence it is smaller than 0.05 after 1.95 seconds, that is, after less than two seconds!

What happens if the billiard ball is observed at various instants of time? For example, if it is observed at times t and $t + 1$, the resulting four dimensional random vector does not approach a uniform distribution for most densities describing initial conditions. Conditional on the ball's initial position, the random vector $(x(t), x(t + 1))$ converges to a uniform distribution as t tends to infinity if and only if the random vector of initial velocities (modulo 1) has a uniform distribution on the unit square (see Example 3 following Proposition 4.3). That is, if $x(t)$ is looked at as a process, dependence on initial conditions does not wash away. As noted in Examples 1 and 2 following Proposition 4.3, there exist various ways of observing the billiard ball's position at various instants of time that do lead to a uniform distribution. For example, if the ball's position is observed at instants that are equally spaced on a logarithmic scale $(t_1 = e^{t+1}, t_2 = e^{t+2}, \ldots, t_k = e^{t+k}, \ldots)$ the resulting infinite dimensional stochastic process does converge (in the weak-star topology, with respect to the cylinder sets) to a distribution uniform on $[0, 1]^\infty$.

4.2.4 Gas Molecules in a Room

Consider a gas molecule moving in a room. Assume the molecule moves in a straight line and that it bounces off the walls, floor and ceiling according to the laws of classical physics (angle of incidence = angle of reflection), without being slowed down. Identify the room with the three dimensional cube $[-\frac{1}{2}, \frac{1}{2}]^3$. If the molecule's initial position and velocity are unknown and described by a joint density, an argument similar to the one carried out in the preceding application shows that the variation distance between the molecule's position at time t and a distribution uniform on $[-\frac{1}{2}, \frac{1}{2}]^3$ is less than c/t where c depends on the total variation of the random variables describing its initial position and velocity.

A more realistic situation is to consider a large number, n, of molecules undergoing uniform rectilinear motion as described in the previous paragraph. Let $v^i = (v_1^i, v_2^i, v_3^i)$ and $x^i(t) = (x_1^i(t), x_2^i(t), x_3^i(t))$ denote the ith particle's initial velocity and position at time t, respectively. Write $v(0)$ and $x(t)$ for the $3n$ dimensional random vectors $(v^1, \ldots, v^n)$ and $(x^1(t), \ldots, x^n(t))$. The dynamical system under consideration is determined by the molecules' initial positions and velocities, $x(0)$ and $v(0)$. Suppose uncertainty about these $6n$ unknown quantities is described by a random vector whose components have bounded mean-conditional variation when conditioned on the remaining coordinates (see Sect. 4.1.2 for the corresponding definition). This holds, in particular, if the joint density of $x(0)$ and $v(0)$ has integrable partial derivatives. If there are no collisions between the molecules, the auxiliary trajectory $y(t) = x(0) + v(0)t$ can be used, just as in Sect. 4.2.3, to show that the variation distance between $x(t)$ and a random vector uniform on $[-\frac{1}{2}, \frac{1}{2}]^{3n}$ is less than c_n/t, where c_n depends on the density describing $(x(0), v(0))$.

Now suppose the particles are allowed to interact. If all molecules have the same mass and collisions are perfectly elastic, every time two particles collide it is if they exchanged identities. That is, at any instant of time the coordinates of $x(t)$ are a permutation (which depends on t) of the coordinates of the corresponding system without interaction. Therefore, if $f(x(t))$ is a measurable function that is invariant under permutations of its argument, Theorem 4.9 and Proposition 2.6 imply that

$$d_V\big(f(x(t)),\, f(U)\big) \le \frac{c_n}{t}, \tag{4.42}$$

where U denotes a distribution uniform on $[-\frac{1}{2}, \frac{1}{2}]^{3n}$.

Consider, for example, the average of the n first coordinates at time t, $\bar{x}_1(t)$. Equation (4.42) implies that, for large values of t, the distribution of $\bar{x}_1(t)$ is near the average of n independent, identically distributed random variables uniform on $[-\frac{1}{2}, \frac{1}{2}]$.

So as to have a limit whose variance does not depend on n, standardize the average first coordinate and consider $\sqrt{12n}\, \bar{x}_1(t)$. Combining (4.42) with the Berry-Esseen Theorem (see e.g. Feller Vol. II, 1971) gives

$$d_K(\sqrt{12n}\, \bar{x}_1(t), Z) \le \frac{c_n}{t} + \frac{3.9}{\sqrt{n}}, \tag{4.43}$$

where d_K denotes the Kolmogorov distance, that is, the supremum over the absolute difference of the distribution functions, and Z is a standard normal density. Hence, for all practical purposes, when n is large enough (which it is in the real world), the standardized average first-coordinate approaches a normal distribution as time passes. A similar argument shows that the three dimensional standardized average position approaches a joint distribution essentially equal to a three dimensional vector composed of independent, identically distributed standard normal densities.

Note that this example gives a justification for the use of the normal distribution from the point of view of the method of arbitrary functions. It also provides a striking example of the quantitative difference between linear and exponential rates of convergence, as is now shown.

Avogadro's Law implies that, at a temperature of 77^0 Farenheit and one atmosphere of pressure, a room 15 feet long, 10 feet wide and 10 feet high has approximately 10^{27} molecules. Imagine the room is initially a vacuum and molecules are allowed to enter. To get an idea of how long it takes for the molecules' positions to distribute approximately uniformly, assume initial positions and velocities are independent and that the latter are independent, identically distributed with total variation equal to V_0. Equation (4.43) implies:

$$d_K(\sqrt{12n}\,\bar{x}_1(t), Z) \leq \frac{10^{27}V_0}{t} + \frac{3.9}{\sqrt{n}}. \tag{4.44}$$

A billion years have less than 10^{17} seconds, so (4.44) does not provide what might be called a useful bound. Yet if initial velocities are a mixture of independent normal distributions with variance larger than one, Remark 2 following Proposition 4.14 and Proposition 2.6.a) imply that

$$d_K(\sqrt{12n}\,\bar{x}_1(t), Z) \leq 10^{27}e^{-\pi^2 t^2/2} + \frac{3.9}{\sqrt{n}},$$

and the Kolmogorov distance is less than 0.05 after 3.7 seconds!

Many other random variables involving the ensemble of molecules in the room may be studied in this way. For example, consider the distribution of the particles nearest to the left and right walls, as viewed from the room's door. These random variables are the largest and smallest order statistics among those describing the molecules' first coordinates. Denote them by $x_{(1)}(t)$ and $x_{(n)}(t)$, respectively. As time passes, the distribution of the normalized random vector $n(x_{(1)}(t) + \frac{1}{2}, \frac{1}{2} - x_{(n)}(t))$ is approximately equal (in the weak-star topology) to two independent exponential random variables with mean equal to one (see Lehmann, 1983, p.395 for the corresponding property of the order statistics of a uniform random variable).

4.2.5 Random Number Generators

Real numbers $x_1, x_2, \ldots$ coming from a physical source are to be used to generate a sequence of "random numbers," $u_1, u_2, \ldots$, that is, numbers whose behavior is like that of a sample of independent, identically distributed random variables uniform on the unit interval. As pointed out by Ripley (1987, p.15), "in practice it seems sufficient to insist that the joint distributions of $(u_{i+1}, \ldots, u_{i+k})$ are not far from uniformity in $[0, 1]^k$ for small values of k (say, $k \leq 6$)."

Assume a physical device is used to obtain a sequence of numbers, $x_1, x_2, \ldots$, that may be assumed to be a realization of some sequence of random variables with unknown joint distribution. In this section Theorem 4.9 is used to construct a new sequence of random variables that is approximately uniform. For a detailed study of a mechanical device used to generate "random numbers" see Inoue, Kumahora, Yoshizawa, Ichimura and Miyitake (1983).

The idea is the following: Assume that $x_1, \ldots, x_n$ originate from a random vector $X = (X_1, \ldots, X_n)'$. Theorem 4.9 implies that the distribution of $(tX)(\mathrm{mod}\,1)$ is approximately uniform if t is sufficiently large and X has a density. If the x_i's are recorded in binary form and t is chosen equal to 2^p for a sufficiently large integer p, no additional calculations are required. To determine how large p should be, standard statistical tests for goodness of fit can be applied, say on sequences of length six, for various values of p. Among those values that pass these tests, it is reasonable to choose the smallest one because this requires less accuracy when the x_i's are recorded. It should be noted that the x_i's need not be independent. All that is needed is that they have a joint density.

4.2.6 Repeated Observations

Consider a block of mass m attached to an ideal spring, free to move over a frictionless horizontal table (see Chap. 1). This time the experiment of stretching the spring and releasing it from rest is conducted many, say n, times.

The spring's constant varies from trial to trial. A series of factors, like temperature, humidity and usage account for this. It therefore makes sense to consider n angular frequencies, $\omega_1, \ldots, \omega_n$, one for each trial, modeled by a joint density, ω, which satisfies the assumptions of Theorem 4.9.

If the spring is released from rest, its displacement from equilibrium at time t on the kth trial is equal to

$$x_k(t) = A_k \cos(\omega_k t),$$

where A_k denotes the corresponding amplitude. Assume the oscillator is observed at time t_k on the kth trial, and denote the n dimensional vector of normalized observations by $\hat{x} = (x_1(t_1)/A_1, \ldots, x_n(t_n)/A_n)$. Theorem 4.9 implies that

$$d_V(\hat{x}, S) \leq \frac{\pi c(\omega, t)}{4}, \tag{4.45}$$

where $S = (S_1, \ldots, S_n)$ is a random vector of independent, identically distributed arcsine laws with density

$$f_S(s) = \frac{1}{\pi\sqrt{1 - s^2}} \qquad ; -1 < s < 1,$$

and

$$c(\omega, t) = \min_{\pi \in S_n} \sum_{i=1}^{n} \frac{V_1\left(\omega_{\pi(i)}, \omega_{\pi(1)}, \ldots, \omega_{\pi(i-1)}\right)}{t_{\pi(i)}}, \tag{4.46}$$

with the V_1's defined in Sect. 4.1.2. Therefore the variation distance between $\hat{x}$ and S can be made arbitrarily small if the instants at which the block's displacements are observed are all large enough.

Let $x = \left(A_1 \cos(\omega_1 t_1), \ldots, A_n \cos(\omega_n t_n)\right)$ denote the vector of original displacements. Just as the angular frequencies, the initial displacements cannot be determined exactly and are therefore described by an n dimensional density. As the initial displacement does not depend on the block's mass or the spring's force constant, this density is independent of ω. Theorem 4.9 and Proposition 2.6 imply that the variation distance between x and $(A_1 S_1, \ldots, A_n S_n)$ is less than or equal to $\pi c(\omega, t)/4$, with $c(\omega, t)$ defined in (4.46) and the S_i's independent of the A_j's. This bound is equal to the one obtained for the normalized displacements (see (4.45)), the main difference being that, in this case, the limit depends on the random vector describing initial displacements. That dependence on initial displacements cannot vanish as time passes is intuitively obvious.

The case where the experiment is performed once and observations are recorded at different instants of time, $t_1, \ldots, t_n$, is now considered. For example, if the oscillator's displacement is observed at times t and $t + k$, the joint distribution of the resulting pair of normalized observations, $\left(\cos(\omega t), \cos(\omega(t + k))\right)$, does not approach a vector of independent arcsine laws as t tends to infinity (see Example 3 following Proposition 4.3). On the other hand, for many schemes where the time interval between successive observations is growing apart as time passes, the corresponding joint distribution is approximately uniform. A particular case is when the oscillator is observed at times t and t^2 (see Example 2 following Proposition 4.3).

In real life the instants at which observations are made are not known exactly and should also be considered random. Theorem 4.9 may still be used in this case to find upper bounds for the variation distance between the vector of normalized displacements and its limiting distribution. The magnitude of these bounds depends on how spread out and smooth the joint density of the time instants $(t_1, \ldots, t_n)$ is. The additional abstraction gained in assuming the t_i's fixed is that it provides a way of making precise the intuitive idea that the position vector of certain dynamical systems approaches a limit that does not depend on the joint density describing initial conditions. This idea, present in Poincaré's work, and fully developed by Hopf, is discussed in detail in the next chapter.

The argument presented in this subsection applies in many situations where an experiment is performed many times, for example, the applications studied in Chap. 3. It is abstracted to a general dynamical system in Chap. 5.

5. Hopf's Approach

Eberhard Hopf was the main developer of the train of thought begun by Poincaré and known as the method of arbitrary functions. Hopf wrote three papers devoted to these developments. This chapter presents and extends his work in a unified fashion.

Sections 1, 2 and 3 are about Hopf's work on dissipative dynamical systems. He showed that if the initial velocity given to a spinning wheel is sufficiently large and frictional forces slowing it down depend only on its velocity, the distribution of its final position is approximately uniform. This result is valid for a wide variety of forces; it assumes uncertainty about the wheel's initial position and velocity is quantified via a joint density (Sect. 5.1).

Hopf also constructed a set of forces for which the distribution of the wheel's final position converges to a distribution that is *not* uniform. The limiting distribution has a very simple expression in terms of the particular force and does not depend on the joint density describing initial conditions. This seems to be the first example where a probability distribution cannot be guessed from symmetry or other plausibility considerations but has to be derived by doing the actual physics.

A general result for n dimensional systems with position coordinates interpreted angularly was proved by Hopf, generalizing work by Copeland (1936). This theorem is reviewed in Sect. 5.2. It applies to dissipative systems slowed down by weak-frictional forces that are a function of only velocity.

Hopf introduced the concept of statistical regularity as a way of making precise the idea of an unpredictable dynamical system. Hopf's results on this subject were developed in the framework of conservative systems. They are extended to arbitrary systems in Sect. 5.4. When the weak-star topology is considered, this generalized concept coincides with that of mixing sequence introduced by Rényi and Révész (1958).

Hopf was aware of the implications his work had on the foundations of probability and mentioned them in every paper he wrote on this subject. He used the method of arbitrary functions to derive the statistical independence of the random variables describing the different coordinates of a dynamical system from physical assumptions. These ideas and the closely related Independence Theorem are reviewed in Sect. 5.5.

Hopf was one of the pioneering researchers in ergodic theory. His work in that area is closely related to the method of arbitrary functions. He introduced the notion of strong-mixing for measure-preserving transformations and showed that it is equivalent to statistical regularity. This result is studied in Sect. 5.6.

The concept of statistical regularity provides a general framework from which most but not all applications discussed in Chaps. 3 and 4 may be viewed. It is extended to that of partial statistical regularity in Sect. 5.7 in order to cover the remaining applications. It is shown that the characterization theorems for statistically regular dynamical systems can be extended to the case of partial statistical regularity.

Hopf used the weak-star topology. In this chapter his results are extended to the variation distance. More importantly, explicit bounds on the rates of convergence are also provided.

Eberhard Hopf was born in Salzburg, Austria in 1902 and died in Bloomington, Indiana in 1983. He attended the University of Berlin where he received his Ph.D. degree in 1925 and his Habilitation in 1929. He visited Harvard University and its Astronomical Observatory in 1930 as a Rockefeller Foundation International Fellow. From 1932 to 1936 he was an Assistant Professor of Mathematics at MIT. In 1936 he succeeded the late professor Lichtenstein at the University of Leipzig. He served there and at the University of Munich until 1947, when he came to New York University as Visiting Professor. He was a faculty member of Indiana University from 1949 onwards, becoming Professor Emeritus in 1972.

On the occasion of Hopf's seventieth birthday, P.M. Anselone (1973) wrote:

"In a remarkably varied and productive career spanning nearly half a century, Eberhard Hopf has made fundamental contributions in several branches of pure and applied mathematics, most notably in partial differential equations, calculus of variations, hydrodynamics, ergodic theory, topological dynamics, integral equations and theoretical astrophysics. His research achievements have considerably advanced our knowledge and understanding of these fields. But, more than that, they have initiated new and vigorous streams of inquiry by many other mathematicians. To a degree seldom matched, Eberhard Hopf has influenced the course of modern mathematical analysis, particularly those areas which have their origins in the physical world.

The mathematical writings and lectures of Professor Hopf are characterized by elegance, keen insight, and technical mastery. In his persistent efforts to get to the heart of the matter, he often perceives far-reaching implications of what had seemed to be special situations..."

Hopf's work on the method of arbitrary functions was, to a large extent, forgotten. A small number of researchers were aware of its philosophical implications (e.g. Savage, 1973 and von Plato, 1983), yet no one thought of further developing the method of arbitrary functions until Persi Diaconis suggested this project to me.

5.1 Force as a Function of Only Velocity: One Dimensional Case

Consider a vertical wheel that spins round a central axis in a clockwise direction and is slowed down by friction. The wheel's position is determined by the angle ϕ a given radius on it makes with a fixed vertical line (see Fig. 5.1).

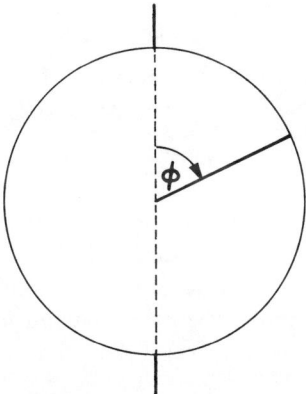

Fig. 5.1. Wheel and angle

Assume frictional forces slowing down the wheel depend only on its velocity. For example, if the only force acting on the wheel is friction with its axle, experiments show that it is approximately constant. Alternatively, if the axle is greased or oiled, the situation is analogous to that of a solid moving slowly in a liquid and friction is approximately proportional to the wheel's velocity (see Barger and Olson, 1973, p.5ff). Newton's equations posit the existence of a function F called force such that

$$\begin{aligned} \omega' &= -F(\omega), \\ \phi' &= \omega, \end{aligned} \tag{5.1}$$

where $\omega(t)$ denotes the wheel's velocity at time t, and ϕ' and ω' the corresponding time derivatives. The minus sign in front of F in (5.1) has been introduced so that $F(\omega) > 0$ for $\omega > 0$. It is assumed that $F(0) = 0$, that is, whatever force opposes the wheel's motion disappears once it stops. This is needed for the wheel to remain at rest once its velocity is zero. It is clear that the wheel's angular velocity does not change signs before it comes to rest, say that $\omega(t) > 0$ for $t < t_1$, where t_1 denotes the time at which it stops.

Consider the following initial conditions:

$$t = 0, \qquad \phi = \phi_0, \qquad \omega = \omega_0.$$

Dividing the first equation in (5.1) by the second one, (strictly speaking, applying the chain rule) leads to

$$\frac{d\phi}{d\omega} = -\frac{\omega}{F(\omega)}.$$

Integrating between $\omega = \omega_0$ and $\omega = 0$ (the assumption $\omega > 0$ for $t < t_1$ is used here):

$$\phi_1 = \phi_0 + \int_0^{\omega_0} \frac{\omega}{F(\omega)} d\omega. \tag{5.2}$$

Assume that, when ω tends to zero, $F(\omega)$ tends to zero more slowly than $\omega^{2-\epsilon}$ for some $\epsilon > 0$, so that the function

$$g(\omega) = \int_0^{\omega} \frac{x}{F(x)} dx \tag{5.3}$$

is well defined. It denotes the distance traveled by a fixed point on the wheel (measured in radians) when its initial velocity is ω. Equation (5.2) may now be written

$$\phi_1 = \phi_0 + g(\omega_0). \tag{5.4}$$

In real life situations, ω_0 and ϕ_0 are not known exactly. Assume uncertainty about them is quantified by specifying a joint density. If this density is sufficiently smooth, the Corollary and Remark 3 following Theorem 3.9, combined with (5.4), give an upper bound for the variation distance between the distribution of the wheel's final position and a uniform distribution on the interval $[0, 2\pi]$. They imply that

$$d_V\left(\phi_1(\text{mod } 2\pi), U\right) \leq \frac{\pi V_1\left(g(\omega_0), \phi_0\right)}{4}, \tag{5.5}$$

where U denotes a distribution uniform on $[0, 2\pi]$, $V_1\left(g(\omega_0), \phi_0\right)$ a measure of how smooth the joint density of $g(\omega_0)$ and ϕ_0 is (see Chap. 4.1), and d_V the variation distance between both random variables, that is, the largest absolute difference among the probabilities they assign to any given event. Given a particular frictional force, (5.3) is used to determine $g(\omega)$ and (5.4) to find an upper bound on the variation distance between the wheel's final position and the corresponding uniform distribution.

Even though the previous paragraph shows how to determine upper bounds for the variation distance between the random variable describing the wheel's final position and a uniform distribution, it does not make precise the intuitive idea that "the wheel's final position is approximately uniform if either its initial velocity is sufficiently large or frictional forces are sufficiently small." Additional abstraction is required to do this. Consider, for example, the case of large initial velocities and just to simplify the argument that follows, assume the initial displacement is fixed and equal to zero. It is not true that if ω_0 has a density that does not take values smaller than some large positive constant M then the variation distance between the wheel's final position and a distribution uniform on $[0, 2\pi]$ is uniformly small. In fact, if the initial velocity has a uniform distribution on $[M, M + \epsilon]$, the variation distance between the corresponding final position and a distribution uniform on $[0, 2\pi]$ can be made arbitrarily close to one by choosing ϵ sufficiently small. In this sense, the wheel's final position does *not* approach a uniform distribution. On the other hand, if ϕ_0 has a distribution uniform on $[0, 2\pi]$ independent of ω_0, Lemma 4.1 can be used to show that the distribution of $\phi(t)(\text{mod } 2\pi)$ is uniform at all times, in particular when the wheel stops, and there is no need for large initial velocities to ensure convergence to a uniform distribution. Of course, there are many practical situations where it does not make sense to assume the wheel's initial position has a distribution uniform on $[0, 2\pi]$, for example, a person placing a record on a turntable often leaves the writing on the record face him or her and all initial positions are not equally likely. Alternatively, the wheel's initial position might be known almost exactly and still its final position should have a distribution approximately uniform on $[0, 2\pi]$ if it is given a sufficiently large initial impulse.

To make mathematically precise the idea that a wheel's final position is approximately uniform if frictional forces are small, Hopf (1936, p.184ff) assumes the shape of the force is fixed and uses a quantity μ called friction coefficient to parametrize its actual size. The force stopping the wheel then is $\mu F(\omega)$ instead of $F(\omega)$, and (5.4) becomes

$$\phi_1 = \phi_0 + \frac{1}{\mu} g(\omega_0).$$

For a fixed joint density describing the wheel's initial position and velocity, ϕ_1 depends only on the friction coefficient μ. Hopf makes precise the idea that the wheel's final position is approximately uniform for weak frictional forces by showing that given any joint density for ω_0 and ϕ_0, the random variable $\phi_1 (\text{mod} \, 2\pi)$ converges to a distribution uniform on $[0, 2\pi]$ as μ tends to zero. Theorem 3.2 implies this is true if convergence in the weak-star topology is considered. That is, the distribution function of $\phi_1 (\text{mod} \, 2\pi)$ evaluated at any x in $[0, 2\pi]$ converges to $x/2\pi$. As is shown in Sect. 5.4, this result remains valid if the stronger notion of variation distance is considered. Further, if the joint density of $g(\omega_0)$ and ϕ_0 is bounded and does not oscillate too much, Theorem 4.9 implies that

$$d_V\left(\phi(\text{mod} \, 2\pi), \, U\right) \; \leq \; \frac{\pi V_1\left(g(\omega_0), \phi_0\right)}{4} \mu, \tag{5.6}$$

and convergence takes place at a rate at least linear in μ.

Hopf (1934, p.98ff) considers two ways of making precise the idea that the wheel's final position is unpredictable for large initial velocities. In the first approach, that will be referred to as the *location case*, he assumes the density describing uncertainty about the initial velocity has a fixed shape that is shifted towards larger values. Let $\omega_0 + a$ denote the initial velocity, where ω_0 is a fixed, non negative density and a a positive real number. Equation (5.4) becomes

$$\phi_1 = \phi_0 + g(\omega_0 + a).$$

First consider the simple case where the force is proportional to the velocity: $F(\omega) = c\omega$ for some positive constant c. Equation (5.4) implies that $\phi_1 = \phi_0 + (\omega_0 + a)/c$ and for most densities describing ω_0, the random variable $\phi_1 (\text{mod} \, 2\pi)$ does not have a limit as a tends to infinity. From (5.3) and (5.4) it is clear that $F(\omega)$ has to grow more slowly than ω if dependence of $\phi_1 (\text{mod} \, 2\pi)$ on initial conditions is to wash away for most densities describing initial conditions. That is, it is necessary to assume

$$\lim_{\omega \to +\infty} \frac{\omega}{F(\omega)} = +\infty,$$

or equivalently

$$\lim_{\omega \to +\infty} \frac{F(\omega)}{\omega} = 0. \tag{5.7}$$

In terms of the function $g(\omega)$ defined in (5.3), condition (5.7) becomes

$$\lim_{\omega \to +\infty} g'(\omega) = +\infty. \tag{5.8}$$

As mentioned at the beginning of this section, if friction is viewed as two solids (wheel and axle) rubbing against each other, it is approximately constant and condition (5.7) holds. Alternatively, if the axle is greased or oiled, friction may be viewed as arising from a solid (the wheel) moving slowly in a liquid (oil or grease) and it therefore is approximately proportional to the wheel's velocity. Assumption (5.7) does not hold in this case.

There exist functions $F(\omega)$ satisfying (5.7) for which, as a tends to infinity, the distribution of $\phi_1(\mathrm{mod}\, 2\pi) = \phi_1(\omega_0, \phi_0, a)(\mathrm{mod}\, 2\pi)$ does not converge to a distribution uniform on $[0, 2\pi]$ for most densities ω_0 (see Sect. 5.4 for a counterexample). The function $\omega/F(\omega)$ cannot oscillate too much as ω tends to infinity if $\phi_1(\omega_0, \phi_0, a)(\mathrm{mod}\, 2\pi)$ is to converge to a uniform distribution as a grows.

Upper bounds for the variation distance between the wheel's final position and a distribution uniform on $[0, 2\pi]$ are now derived. The argument is conditional on the value of the wheel's initial position, ϕ_0. Proposition 2.8 extends the bounds to the general case. To motivate the additional assumption required for $F(\omega)$, note that under the assumptions made so far, (5.5) becomes

$$d_V\big(\phi_1(\mathrm{mod}\, 2\pi),\, U\big) \leq \frac{\pi}{4} V\big(g(\omega_0 + a)\big),$$

where $V\big(g(\omega_0 + a)\big)$ denotes the total variation of the corresponding distribution (see Chap. 3.1). Assume the density of ω_0, $f(\omega)$, is continuous and piecewise differentiable. As $\omega_0 \geq 0$ it follows that $f(0) = 0$. Proposition 3.8.d) can then be used to find an upper bound for $V\big(g(\omega_0 + a)\big)$ leading to

$$d_V\big(\phi_1(\omega_0, a)(\mathrm{mod}\, 2\pi),\, U\big) \leq \frac{\pi}{4} \bigg\{ \int_0^{+\infty} \frac{|f'(\omega)|}{g'(\omega + a)}\, d\omega$$
$$+ \int_0^{+\infty} \frac{|g''(\omega + a)|}{|g'(\omega + a)|^2} f(\omega)\, d\omega \bigg\}, \tag{5.9}$$

where $g(\omega_0)$ denotes the distance the wheel travels before it stops (see (5.3) and (5.4)). Due to (5.8) the first term on the right hand side of (5.9) tends to zero as a tends to infinity if $f(\omega)$ has bounded variation, that is, if $\int |f'(\omega)|d\omega < +\infty$.

Additional conditions are needed to ensure that the second term on the right hand side of (5.9) also tends to zero as a tends to infinity. The conditions under which Hopf proved his results are compared with those derived below at the end of this section.

If either

$$\lim_{a \to \infty} \sup_{x \geq a} \frac{|g''(x)|}{|g'(x)|^2} = 0, \tag{5.10}$$

or $f(\omega)$ is bounded and

$$\int_M^\infty \frac{|g''(\omega)|}{|g'(\omega)|^2} d\omega \ < +\infty \quad \text{for some real number M,} \tag{5.11}$$

the second term on the right hand side of (5.9) tends to zero. It follows that the random variable $\phi_1(\omega_0, a)(\mathrm{mod}\, 2\pi)$ converges, in the variation distance, to a distribution uniform on $[0, 2\pi]$ if $g(\omega)$ satisfies conditions (5.8) and (5.11) and the density of ω_0 is continuous, piecewise differentiable and bounded. If condition (5.11) is replaced by (5.10), the conclusion remains valid if the density of $f(\omega)$ is not bounded. In Sect. 5.4 it is shown that $\phi_1(\mathrm{mod}\, 2\pi)$ converges in the variation distance to a uniform distribution for any density describing ω_0. The rates of convergence that can be derived from (5.9) do not apply to every density (Proposition 3.7 may be used to construct a counterexample).

If, for sufficiently large ω, $g(\omega)$ is convex, then $|g''(\omega)|/|g'(\omega)|^2$ is equal to the derivative of $-1/g'(\omega)$ and this can be used to derive (5.11) from (5.8) in this case.

Bounds that are larger than those provided by (5.9), but usually easier to calculate, are given by

$$d_V\big(\phi_1(\omega_0, a)(\mathrm{mod}\, 2\pi),\ U\big) \ \leq\ \frac{\pi}{4} \left\{ \frac{V(\omega_0)}{\min_{x \geq a} g'(x)} + \sup_{x \geq a} \frac{|g''(x)|}{|g'(x)|^2} \right\}, \tag{5.12}$$

where $V(\omega_0)$ denotes the total variation of ω_0.

The bounds in (5.12) and (5.9) are conditional on the value of the initial position ϕ_0. Proposition 2.8 can be used to extend them to the general case. For example, (5.12) becomes

$$d_V\big(\phi_1(\omega_0, a)(\mathrm{mod}\, 2\pi),\ U\big) \ \leq\ \frac{\pi}{4} \left\{ \frac{V_1(\omega_0, \phi_0)}{\min_{x \geq a} g'(x)} + \sup_{x \geq a} \frac{|g''(x)|}{|g'(x)|^2} \right\}, \tag{5.13}$$

where $V_1(\omega_0, \phi_0)$ denotes the mean-conditional variation of ω_0 with respect to ϕ_0 (see Chap. 4.1). Of course, if ϕ_0 and ω_0 are independent, (5.13) simplifies to (5.12).

A family of frictional forces that quite naturally models the case where the wheel is viewed somewhere between a solid rubbing against another solid ($F(\omega) = c$) and a solid moving slowly in a liquid ($F(\omega) = c\omega$) is provided by the power family: $F(\omega) = c\omega^\gamma$, $0 \leq \gamma \leq 1$. Equation (5.13) then implies

$$d_V\big(\phi_1(\omega_0, a)(\mathrm{mod}\, 2\pi),\ U\big) \ \leq\ \frac{\pi c}{4a^{1-\gamma}} \left(V_1(\omega_0, \phi_0) + \frac{1-\gamma}{a} \right), \tag{5.14}$$

and $\phi_1(\mathrm{mod}\, 2\pi)$ converges to a distribution uniform on $[0, 2\pi]$ as a tends to infinity, if $\gamma < 1$. The bound in (5.14) tends to zero very slowly if γ is near one. This is consistent with the fact that, as pointed out before, there usually is no convergence if $\gamma = 1$.

Equation (5.14) is now used to obtain numeric upper bounds for the variation distance between a carnival wheel's final position and a distribution uniform on $[0, 2\pi]$. A carnival wheel is not only slowed down by friction with its axle but also by a rubber or leather pointer that bumps noisily against nails arranged on the wheel's rim. The fact that the force slowing down the wheel also depends on its position is taken into account in Sect. 5.3. Assume the pointer has been removed and the frictional force encountered by the wheel is approximately constant: $F(\omega) = c$, that is, the axle is neither greased nor oiled.

Uncertainty about the wheel's initial velocity does usually not depend much on its initial position. Therefore the total variation of ω_0 conditioned on a specific value of ϕ_0 is (approximately) the same for all possible values of ϕ_0 and $V_1(\omega_0, \phi_0)$ is equal to the total variation of ω_0, $V(\omega_0)$ (see Chap. 3.1). To use the bounds obtained in (5.14), some idea about typical values of the constant c and the initial velocity ω_0 are needed. It is hard to measure these quantities directly, yet anyone who has visited an entertainment park has some idea about the number of times a carnival wheel rotates, n, and the time it takes before it stops, t_1. To see how the latter quantities determine ω_0 and c, first note that integrating the first equation in (5.1) between $t = 0$ and $t = t_1$ shows that $\omega_0 = ct_1$. Equation (5.2) leads to $2\pi n = \omega_0^2/2c$. Hence

$$\omega_0 = \frac{4\pi n}{t_1}; \qquad c = \frac{4\pi n}{t_1^2}. \qquad (5.15)$$

My personal recollection is that it takes between one and two minutes before a carnival wheel stops spinning and the number of cycles it completes is somewhere between seven and ten. Equation (5.15) then implies that the initial velocity lies between $7\pi/30$ and $2\pi/3$ radians per second. Assume ω_0 is a mixture of unimodal densities supported by $[7\pi/30, +\infty)$ and bounded by twice the maximum value of a distribution uniform on $[7\pi/30, 2\pi/3]$. Proposition 3.8 then leads to

$$V(\omega_0) \leq \frac{120}{13\pi}.$$

A similar argument shows that c may be bounded from above by $\pi/90$. The assumptions made to obtain an upper bound for $V(\omega_0)$ imply that $a \geq 7\pi/30$. Equation (5.14) with $\gamma = 0$ now leads to

$$d_V\left(\phi_1(\omega_0, a)(\mathrm{mod}\, 2\pi),\, U\right) \leq 0.16. \qquad (5.16)$$

One piece of information that has not been taken into account so far is that usually the initial position and velocity may be assumed independent. Proposition 3.12 then leads to

$$d_V\left(\phi_1(\omega_0, a)(\mathrm{mod}\, 2\pi),\, U\right) \leq d_V(\phi_0,\, U). \qquad (5.17)$$

Therefore, if a hypothetical carnival wheel is considered and no information about its initial position is available, (5.17) should lead to very good bounds because the distribution of ϕ_0 is approximately uniform on $[0, 2\pi]$. On the other hand, if someone at an entertainment park wants to make an assessment about where the carnival wheel will land the next time it is spun and he or she knows where it landed the last time, (5.16) provides better bounds than (5.17). The bound in (5.16) becomes larger if more data about where the wheel started and landed on previous occasions is available, because this leads to sharper densities for ω_0 and therefore a larger total variation. Therefore it is quite plausible that one could make money at a carnival wheel. For a story of how some scientists tried to make money playing roulette, using the fact that all possible slots where the roulette ball might land are *not* equally likely if enough information about initial conditions is available, see Bass (1986).

A second approach considered by Hopf (1934, p.100) for large initial velocities is the *scalar case*. It assumes the shape of the density of the initial velocity fixed except for a scaling factor b, that is, the wheel's initial velocity is distributed like $b w_0$ with w_0 fixed. The idea is to find conditions on the frictional force $F(\omega)$ that ensure convergence of the random variable $\phi_1 (\bmod 2\pi)$ to a uniform distribution, as b tends to infinity, for any joint density describing initial conditions.

First assume that the wheel's initial position is known. An argument similar to the one given in the location case implies that, if w_0 has a continuous, piecewise differentiable density supported by $[m, M]$ then

$$
d_V\left(\phi_1(w_0, b)(\bmod 2\pi), U\right) \leq \frac{\pi}{4}\left\{\int_m^M \frac{|f'(\omega)|}{bg'(b\omega)}d\omega + \int_m^M \frac{|g''(b\omega)|}{|g'(b\omega)|^2}f(\omega)d\omega\right\}. \quad (5.18)
$$

It follows that sufficient conditions for convergence as b tends to infinity are

$$
\lim_{b\to+\infty} \min_{bm\leq\omega\leq bM} bg'(\omega) = +\infty, \quad (5.19)
$$

and

$$
\lim_{b\to+\infty} \sup_{bm\leq\omega\leq bM} \frac{|g''(\omega)|}{|g'(\omega)|^2} = 0. \quad (5.20)
$$

In Sect. 5.4 it is shown that conditions (5.19) and (5.20) are sufficient for convergence of $\phi_1(w_0, b)(\bmod 2\pi)$ to a distribution uniform on $[0, 2\pi]$ for any joint density describing w_0 and ϕ_0. The additional smoothness assumptions made on the density of w_0 are only needed to derive bounds like (5.18). These bounds are useful only if w_0 is bounded away from 0, that is, if $m > 0$. This is not surprising, for otherwise the random variable describing how far the wheel travels when its initial velocity is $b w_0$, $g(b w_0)$, may take values arbitrarily close to zero no matter how large the scaling factor b is.

Bounds that are larger than those provided by (5.18), but usually easier to calculate, are given by

$$
d_V\left(\phi_1(w_0, b)(\bmod 2\pi), U\right) \leq \frac{\pi}{4}\left(\frac{V(w_0)}{b\min_{bm\leq\omega\leq Bm} g'(\omega)}\right.
$$
$$
\left. + \sup_{bm\leq\omega\leq bM} \frac{|g''(\omega)|}{|g'(\omega)|^2}\right).
$$

If the sign of the second derivative of $g(\omega)$ does not change for sufficiently large ω, that is, if $g(\omega)$ eventually becomes convex or concave, conditions (5.19) and (5.20) simplify to

$$
\lim_{\omega\to+\infty} \frac{F(\omega)}{\omega^2} = 0 \qquad \text{and} \qquad \lim_{\omega\to+\infty} \frac{F'(\omega)}{\omega} = 0.
$$

Note that in the location case the fact that the derivative of $g(\omega)$ tends to infinity implies that $g(\omega)$ cannot be concave.

Frictional forces acting on a body moving very fast in a liquid (e.g. an airplane or a baseball) are approximately proportional to the square of the body's velocity. Though rotating wheels do not spin that fast, a family of forces that considers initial velocities that are neither too large nor too small is provided by $F(\omega) = c\omega^\gamma$ with $0 \leq \gamma < 2$. Then (5.18) implies

$$d_V\left(\phi_1(\omega_0, b)(\text{mod } 2\pi), U\right) \leq \begin{cases} \frac{\pi}{4cm^{1-\gamma}b^{2-\gamma}}\left(V(\omega_0) + \frac{1-\gamma}{m}\right) & \text{if } 0 < \gamma \leq 1; \\[2ex] \frac{\pi}{4cb^{2-\gamma}}\left(V(\omega)M^{\gamma-1} + \frac{\gamma-1}{m^{2-\gamma}}\right) & \text{if } 1 < \gamma < 2. \end{cases}$$

Therefore, the distribution of the wheel's final position converges to a distribution uniform on $[0, 2\pi]$ if $\gamma < 2$. In particular, in contrast to the location case, when friction is proportional to the wheel's velocity, $\phi_1(\text{mod } 2\pi)$ does converge to a uniform distribution in the scalar case. A calculation from first principles based on (5.1) and (5.4) shows that if friction is proportional to the square of the velocity when velocities are large, then for most densities ω_0 the distribution of $\phi_1(\omega_0, b)$ does not have a limit as b tends to infinity.

This approach may be generalized in many ways. For example, a combination of the location and scalar cases might be considered (much in the spirit of a Poisson random variable). Alternatively, the scalar approach can be applied to the logarithm of ω_0. This is equivalent to fixing ω_0 and looking at the distribution of the final position associated to ω_0^γ as γ tends to infinity. In this case $\phi_1(\omega_0, \gamma)(\text{mod } 2\pi)$ does converge to a distribution uniform on $[0, 2\pi]$ when, for large velocities, friction is proportional to the square of the velocity. In fact, if $\log \omega_0$ has bounded variation, Proposition 3.8.d) implies that

$$d_V\left(\phi_1(\omega_0^\gamma)(\text{mod } 2\pi), U\right) \leq \frac{\pi V(\log \omega_0)}{4\gamma}.$$

Before concluding this section, the hypothesis under which Hopf (1934, p.98ff) showed convergence in the weak-star topology are compared with those used in this section to prove convergence in the variation distance (and therefore in the weak-star topology). In the location case, Hopf showed that $\phi_1(\text{mod } 2\pi)$ converges to a distribution uniform on $[0, 2\pi]$ in the weak-star topology if the function $g(\omega_0)$ describing the distance traveled by the wheel when its initial velocity is ω_0 (see (5.3)) satisfies

$$\lim_{\omega \to +\infty} g'(\omega) = +\infty \tag{5.21}$$

and

$$\lim_{\omega \to +\infty} \left| \frac{g'(\omega + s)}{g'(\omega)} - 1 \right| = 0, \tag{5.22}$$

where (5.22) holds uniformly in every finite interval containing s.

In this section it was shown that convergence in the variation distance (and therefore in the weak-star topology) follows from (5.21) and either

$$\int_M^{+\infty} \frac{|g''(\omega)|}{|g'(\omega)|^2}\, d\omega < +\infty \quad \text{for some } M > 0, \tag{5.23}$$

or

$$\lim_{a \to \infty} \sup_{x \geq a} \frac{|g''(x)|}{|g'(x)|^2} = 0. \tag{5.24}$$

If $g(\omega)$ is convex for sufficiently large ω, these conditions are more general than Hopf's because convexity of $g(\omega)$ together with (5.21) imply (5.22). For example, $F(\omega) = \omega e^{-\omega}$ satisfies (5.21), (5.23) and (5.24) but not (5.22). Yet if $g(\omega)$ is not eventually convex, both sets of conditions are not comparable: an example satisfying (5.21) and (5.22) but neither (5.23) nor (5.24) is given by $g(x) = 1 + \int_0^x \left(x + \sin(e^x) \right) dx$.

The conditions under which Hopf (1934, p.100) proved convergence in the scalar case, namely

$$\lim_{\omega \to +\infty} \omega g'(\omega) = +\infty \qquad \text{and} \qquad \lim_{\omega \to +\infty} \omega \frac{g''(\omega)}{g'(\omega)} = 0,$$

seem unnecessarily restrictive. Many forces for which it would seem natural that the method of arbitrary functions applies do not satisfy them. For example, if for large velocities $F(\omega) = c\omega^\gamma$, Hopf's conditions are only satisfied if $\gamma = 1$ while those derived in this section apply for $0 \le \gamma < 2$.

5.2 Force as a Function of Only Velocity: Higher Dimensions

In this section, Hopf's work for dissipative systems whose position (and velocity) is described by many variables is reviewed. Hopf (1937 p.165ff) considered the case of weak frictional forces that only depend on the system's velocity. He assumed uncertainty about initial conditions is quantified using a joint density and that position variables are interpreted angularly. He proved that the random vector describing the system's final position converges, in the weak-star topology, to a uniform distribution as the friction coefficient, μ, tends to zero. In this section convergence in the variation distance is proved. It is also shown that if the density describing initial conditions is sufficiently smooth, convergence is linear in μ.

To begin, here is a motivating example showing the use of a higher dimensional analysis. In 1737 Georges-Louis Leclerc, Comte de Buffon, wrote on the *clean tile* game: In a room tiled or paved with equal tiles, a coin is thrown upwards; one of the players bets that after its fall the coin will rest cleanly, that is, on one tile only; the second bets that the coin will rest on two tiles, that is, will cover one of the cracks which separate them and so on. It is required to find the chance of winning for each player.

As a special case, Buffon assumed the room is covered with parallel lines and a needle is thrown instead of a coin. He showed that, provided the length of the needle, l, is less than the distance between the lines, δ, the probability of an intersection is $p = 2l/\pi\delta$. This problem, known as Buffon's needle problem, began the discipline of Geometric Probability.

The needle's final position is described using two variables: the distance of its center of gravity to the nearest line to its left, $x(\bmod \delta)$, and the angle it makes with the parallel lines, $\theta(\bmod 2\pi)$ (see Fig. 5.2).

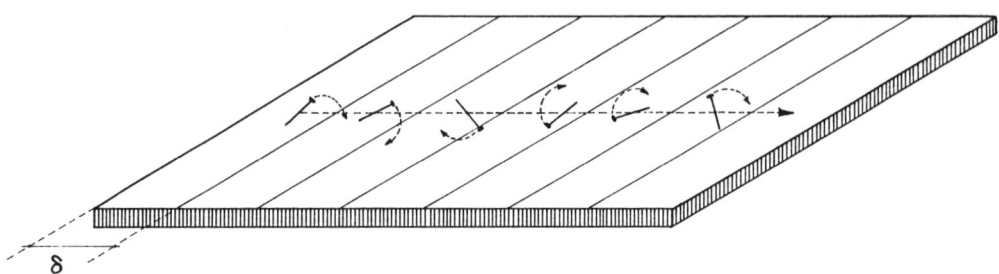

Fig. 5.2. Buffon's needle

Buffon's calculations assume that $(x(\mathrm{mod}\,\delta), \theta(\mathrm{mod}\,2\pi))$ has a distribution uniform on $[0,\delta] \times [0,2\pi]$. Hopf (1934, p.100ff; 1936, p.184ff and 1937, p.162) applied the method of arbitrary functions to derive this assumption. In Hopf's version, the needle is initially on the floor. An impulse is given to it so that it moves rotating and translating at the same time until it stops. Newton's equations are:

$$
\begin{aligned}
Mx'' &= N, \\
I\theta'' &= T,
\end{aligned}
\tag{5.25}
$$

where M denotes the needle's mass, I its moment of inertia, N the frictional force opposing its rectilinear motion and T the corresponding torque. Assume N and T only depend on the needle's linear and angular velocities. So as to consider the case of weak frictional forces, N and T are replaced by μN and μT and the distribution of the needle's final position, $(x_1(\mathrm{mod}\,\delta), \theta_1(\mathrm{mod}\,2\pi))$, as the friction coefficient μ tends to zero, is studied.

In the general case, where n coordinates determine the system's position, Newton's equations lead to

$$
\begin{aligned}
\omega'(t) &= \mu F\big(\omega(t)\big), \\
\phi'(t) &= \omega(t),
\end{aligned}
\tag{5.26}
$$

where $\phi(t) = (\phi_1(t), \ldots, \phi_n(t))$ denotes the needle's position at time t, $\omega(t)$ the vector of corresponding velocities and $\phi'(t)$ and $\omega'(t)$ their time derivatives. The position vector $\phi(t)$ may take any value in IR^n yet its coordinates are interpreted angularly. For example, Buffon's needle may travel very far (and therefore the value of x may be very large) yet our interest centers on the needle's distance to the nearest line it just passed, $x(\mathrm{mod}\,\delta)$, and this is a number between 0 and δ. In the general case, therefore, our attention centers on $\phi(t)(\mathrm{mod}\,1)$. The fact that $a(\mathrm{mod}\,\delta)/\delta = (a/\delta)(\mathrm{mod}\,1)$ implies that the results that follow are easily extended to deal with $(\phi_1(\mathrm{mod}\,\delta_1), \ldots, \phi_n(\mathrm{mod}\,\delta_n))$.

Consider the following set of initial conditions for (5.26):

$$
t = 0; \qquad \phi = \phi_0; \qquad \omega = \omega_0.
$$

In this version of the problem the force does not depend on the system's position. Hopf (1937, p.165) found a change of variable that leads to a *new system of differential equations* that *does not* involve the friction coefficient μ. Of course *the change of variable does involve* μ. It is

$$\mu t = \tau; \qquad u(\tau) = \omega(t); \qquad x(\tau) = \mu\big(\phi(t) - \phi(0)\big).$$

Equation (5.26) becomes

$$\begin{aligned} u' &= F(u), \\ x' &= u, \end{aligned} \tag{5.27}$$

with initial conditions

$$\tau = 0; \qquad x = 0; \qquad u = \omega_0. \tag{5.28}$$

Denote by $\tau_1 \leq +\infty$ the time at which the system described by (5.27) and (5.28) comes to rest. Then

$$x(\tau_1) = \int_0^{\tau_1(\omega_0)} u(\tau, \omega_0) d\tau, \tag{5.29}$$

where the notation introduced stresses the fact that τ_1 and $u(t)$ depend on not only on the force F but also on ω_0. What matters is that they do not depend on the friction coefficient μ.

As $x(\tau_1)$ is equal to $\mu(\phi_1 - \phi_0)$, (5.29) implies that

$$\phi_1 = \phi_0 + \frac{1}{\mu} Y(\omega_0), \tag{5.30}$$

where ϕ_1 denotes the system's final position (not to be confused with the first coordinate of the position vector ϕ) and the i-th coordinate of $Y(\omega_0)$ is equal to

$$Y_i(\omega_0) = \int_0^{\tau_1(\omega_0)} u_i(\tau, \omega_0) d\tau. \tag{5.31}$$

Assume uncertainty about the system's initial position is quantified using a joint density for ϕ_0 and ω_0. Theorems 4.2 and 4.4 applied to (5.30) implies that $\phi_1(\mathrm{mod}\,1)$ converges in the weak-star topology to a distribution uniform on $[0,1]^n$ that is independent of initial conditions, as μ tends to zero. In Sect. 5.4 it is shown that convergence still takes place for every density if the stronger notion of variation distance is considered. Therefore the final position of a dissipative dynamical system with coordinates interpreted angularly and slowed down by forces that only depend on its velocity is approximately uniform for weak frictional forces.

If the density of $Y(\omega_0)$ is sufficiently smooth, the Corollary following Theorem 4.9 applied to (5.30) implies that

$$\mathrm{d}_V\big(\phi_1(\mathrm{mod}\,1), U_n\big) \leq \frac{S_1\big(Y(\omega_0), \phi_0\big)}{8} \mu, \tag{5.32}$$

where $S_1\big(Y(\omega_0), \phi_0\big)$ is a measure of smoothness of the density of $\big(Y(\omega_0), \phi_0\big)$. Convergence takes place at a rate at least linear in μ. The discussion following Theorem

4.9 implies that the bound in (5.32) is asymptotically sharp – in the sense that there exist densities for which it can be approximated arbitrarily closely – if given any n dimensional rectangle $\mathcal{R}$ there exists a density for w_0 such that $Y(w_0)$ has a distribution uniform on $\mathcal{R}$. That this is quite generally the case follows from Theorem 20.3 in Hewitt and Stromberg (1965).

To apply the bounds derived in (5.32), the explicit functional form of $Y(w_0)$ is needed. As $Y(w_0)$ depends on the solution of the system of differential equations defined in (5.27), it is usually not possible to obtain this expression explicitly. Hopf (1934, p.101 and 1936, p.186) noted that an interesting exception is when friction is proportional to velocity. In Buffon's needle example, assume the needle is moving on a very slippery surface (e.g. the bottom of a swimming pool) and that the frictional force opposing its rectilinear motion is approximately proportional to its linear velocity, v, while the torque opposing its circular motion is proportional to its angular velocity, β. Newton's equations (see (5.25)) result in

$$M x'' = \mu x' = \mu v,$$
$$I \theta'' = \mu \theta' = \mu \beta,$$

where M denotes the needle's mass, I its moment of inertia and μ the constant of proportionality. To simplify the expressions that follow, assume units are chosen so that $I = M = 1$. Then the vector $Y(w_0)$ introduced in (5.31) is equal to (v_0, β_0), where v_0 and w_0 denote the initial linear and angular velocities, respectively. Theorem 4.9 implies that

$$d_V\left((x_1(\mathrm{mod}\,\delta), \theta_1(\mathrm{mod}\,2\pi)), U\right) \leq \frac{\pi S}{4}\mu,$$

where S denotes a measure of smoothness of the joint density of v_0, β_0, x_0 and θ_0 (see Chap. 4.1) and U is a distribution uniform on $[0, \delta] \times [0, 2\pi]$. For example, if all four initial conditions are assumed independent, S is equal to the sum of the total variation of v_0 and β_0.

This section concludes with some remarks on Hopf's proof of weak-star convergence of $\phi_1(w_0, \mu)(\mathrm{mod}\,1)$ to a distribution uniform on $[0, 1]^n$ (see Hopf, 1937, p.165ff). Relying strongly on properties of differential equations he showed that it suffices to prove that

$$\lim_{a \to 0} \frac{1}{a} \int_0^a h\left(\frac{y_1(t)}{\mu}, \dots, \frac{y_n(t)}{\mu}\right) dt = \int_{[0,1]^n} h(x_1, \dots, x_n) dx_1 \dots dx_n \qquad (5.33)$$

for any continuous function $h(x_1, \dots, x_n)$ that is periodic in every component (with period one) and any function $y(t) = (y_1(t), \dots, y_n(t))$ satisfying certain smoothness conditions. Equation (5.33) has the flavor of an ergodic theorem: the left hand side is a time average while the right hand side is a phase space average.

Hopf's proof of (5.33) is based on approximating $h(x_1, \dots, x_n)$ by a product of complex exponentials. Once this is done, the proof reduces to the Riemann-Lebesgue Lemma (see Billingsley, 1986, p.354).

Copeland (1936) proved the same theorem under the additional assumption that the transformation $(x_0, u_0) \to (x_1, u_0)$ defined implicitly by (5.28) and (5.29) has a continuous, non vanishing Jacobian.

The proof given in this section seems conceptually simpler and proves a stronger result: convergence in variation distance. Further, assuming initial densities are sufficiently
smooth, convergence was shown to be linear in μ.

5.3 The Force Also Depends on the Position

One of Hopf's main results was constructing a physically meaningful problem where
the method of arbitrary functions applies but the limiting distribution is not uniform
(see Hopf, 1936, p.187ff). He showed that, given any density on the interval $[0, 2\pi]$,
there exists a frictional force slowing down a spinning wheel such that its final position
approaches the given density as either the initial velocity grows or the friction coefficient
becomes small.

Consider a carnival wheel, this time taking into account the leather pointer that
bumps against the nails arranged on the wheel's rim. The wheel's position is determined
by the angle ϕ a given radius on it makes with a fixed vertical line (see Fig. 5.1). Due
to the leather pointer and the nails, the force stopping the wheel does depend on its
position: it is larger in the neighborhood of a nail. One way of combining this with the
fact that friction also might depend on the wheel's velocity is to assume that for a given
velocity, friction is proportional to a fixed function of the position. The force $F(\phi, \omega)$
slowing down the wheel then becomes

$$F(\phi, \omega) \ = \ r(\omega)\chi(\phi), \tag{5.34}$$

where ω and ϕ denote the wheel's velocity and position, respectively. The function $\chi(\phi)$
for a frictional force corresponding to nails around the rim should approximately look
like Fig. 5.3 with the peaks at the points where the nails are located. By eventually
rescaling $r(\omega)$ there is no loss of generality in assuming

$$\int_0^{2\pi} \chi(\phi)d\phi \ = \ 2\pi. \tag{5.35}$$

The wheel stops when its velocity, ω, is zero and therefore the condition $r(0) = 0$ is
imposed so that it stays at rest. It is also clear that $\chi(\phi)$ is periodic, with period 2π,
and that the wheel's velocity does not change sign before it stops.

Newton's equations are

$$\begin{aligned} \omega' &= -r(\omega)\chi(\phi), \\ \phi' &= \omega, \end{aligned} \tag{5.36}$$

with initial conditions

$$t = 0 ; \qquad \phi = \phi_0 ; \qquad \omega = \omega_0.$$

Let ϕ_1 denote the wheel's final position and assume its initial position and velocity
are described by a joint density. Does there exist a distribution on $[0, 2\pi]$ to which
$\phi_1 (\mathrm{mod}\, 2\pi)$ converges as the initial velocity increases or frictional forces weaken? If such
a limiting distribution exists, it is not uniform because the wheel has a larger probability
of stopping near one of the nails than elsewhere. In the case where friction did not depend

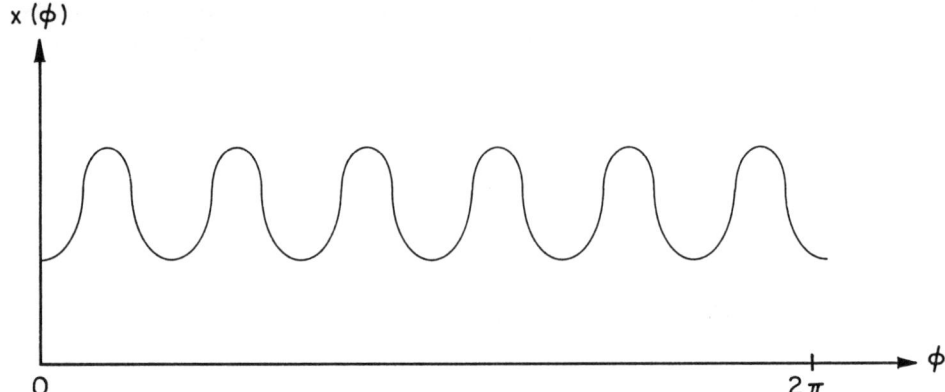

Fig. 5.3. Force component that depends on the wheel's position: symmetric case

on the wheel's position (see Sect. 5.1) symmetry considerations show that if there exists a limiting distribution that does not depend on initial conditions it has to be uniform on $[0, 2\pi]$. There are no symmetry conditions that are useful in this case. Hopf (1936, p.179) pointed out that *"this is the first example where probability cannot be guessed by symmetry or other plausibility considerations."* The power of Hopf's ingenious approach can be fully appreciated in this problem.

First consider the case of weak frictional forces. A friction coefficient μ is introduced on the right hand side of the first equation in (5.36) and the distribution of $\phi_1(\mathrm{mod}\, 2\pi)$ as a function of μ, for a fixed density (ϕ_0, ω_0) describing initial conditions, is studied as μ tends to zero. If there exists a limit that does not depend on initial conditions it makes sense to say that there exists a universal limiting distribution, otherwise not.

Proceeding the same way as in Sect. 5.1 leads to

$$\int_0^{\omega_0} \frac{\omega}{r(\omega)} d\omega = \int_{\phi_0}^{\phi_1} \chi(\phi) d\phi, \tag{5.37}$$

where ϕ_0 and ϕ_1 denote the wheel's initial and final positions, respectively, and ω_o its initial velocity.

Let

$$g(\omega_0) = \int_0^{\omega_0} \frac{\omega}{r(\omega)} d\omega. \tag{5.38}$$

Using (5.35) and the periodicity of $\chi(\phi)$ leads to

$$\left\{ \int_0^\phi \chi(\phi) d\phi \right\} (\mathrm{mod}\, 2\pi) = \int_0^{\phi(\mathrm{mod}\, 2\pi)} \chi(\phi) d\phi, \tag{5.39}$$

and hence the auxiliary variable

$$\psi(\phi) = \int_0^\phi \chi(\phi) d\phi \tag{5.40}$$

may also be interpreted angularly, that is, $\psi(\phi(\bmod 2\pi)) = \psi(\phi)(\bmod 2\pi)$. Denoting by ψ_0 and ψ_1 the initial and final values of ψ_1, (5.37), (5.38) and (5.39) imply that

$$\psi_1 = \psi_0 + g(\omega_0).$$

Comparing this equation with (5.4) shows that any conclusion valid for $\phi_1(\bmod 2\pi)$ when the force only depends on the wheel's velocity, holds for ψ_1 in the role of ϕ_1 in the general case. For example, if the force is equal to $\mu r(\omega)\chi(\phi)$, the distribution of $\psi_1(\bmod 2\pi)$ converges in the variation distance to a distribution uniform on $[0, 2\pi]$ for any joint density describing the wheel's initial position and velocity. Equations (5.39) and (5.40), the change of variable formula and Proposition 2.6 imply that $\phi_1(\bmod 2\pi)$ converges, in the variation distance, to a distribution supported by $[0, 2\pi]$ with density equal to $\chi(\phi)$ (see Fig. 5.3). Even though this density is not uniform, the probability it assigns to every slot, that is, the space between two consecutive nails on the wheel's rim, is approximately the same if the nails are equally spaced.

Consider a carnival wheel where the distance between consecutive nails varies. Assume, for example, that the arc length between consecutive nails is either once, twice or three times a standard length. The argument given above implies that the density describing the wheel's final position looks approximately like Fig. 5.4.

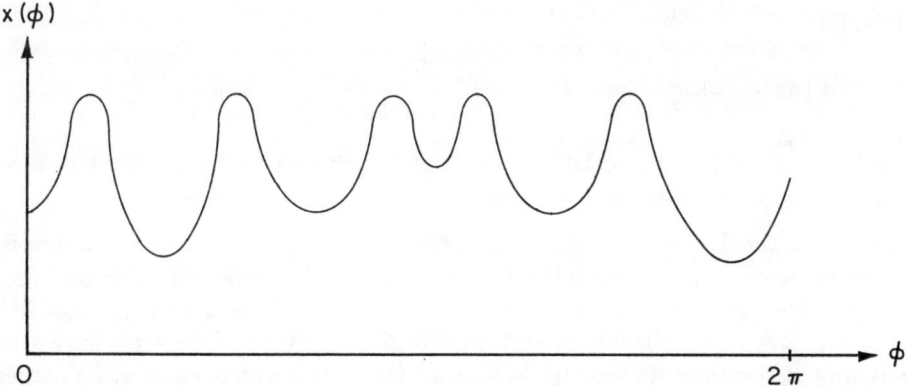

Fig. 5.4. Force component that depends on the wheel's position: asymmetric case

It is now easy to see that if prizes are inversely proportional to slot size, our best bet is to play on the smaller slots. The fair price for such a game is not inversely proportional to the arc length between successive nails. Alternatively assume all slots have the same size but the number of nails per slot differs. Then those slots with more nails should win more often.

It should also be noted that the problem of estimating the part of the frictional force that depends on the wheel's position, that is $\chi(\phi)$, is equivalent to estimating the density of the limiting distribution. This statement relies heavily on the assumption made regarding the way the components of friction due to the wheel's position and velocity combine (see (5.34)).

To obtain upper bounds for the variation distance between $\phi_1(\mathrm{mod}\,2\pi)$ and the limiting distribution, L, in the case of weak frictional forces, note that Proposition 2.6 and the fact that $\phi(\mathrm{mod}\,2\pi)$ is a function of $\psi(\mathrm{mod}\,2\pi)$ combined with (5.6) imply that if the joint density of $g(\omega_0)$ and ϕ_0 is sufficiently smooth,

$$d_V(\phi(\mathrm{mod}\,2\pi),\,L) \;\leq\; \frac{\pi V_1\big(g(\omega_0),\phi_0\big)}{4}\mu,$$

and convergence is at least linear in μ.

Consider the case of large initial velocities. In the location case, where the initial velocity is equal to the sum of a fixed positive density ω_0 and a positive real number a, the distribution of the carnival wheel's final position converges in the variation distance to L if the function $g(\omega)$ defined in (5.38) satisfies

$$\lim_{\omega\to+\infty} g'(\omega) \;=\; +\infty$$

and

$$\int_M^{+\infty} \frac{|g''(\omega)|}{|g'(\omega)|^2}\,d\omega \;<\; +\infty \quad \text{for some } M > 0.$$

Further, if the density of ω_0 has bounded mean-conditional variation when conditioned on ϕ_0 (see Chap. 4.1 for the definition) and the joint density of ω_0 and ϕ_0 is continuous, piecewise differentiable, then

$$d_V\big(\phi_1(\omega_0,a)(\mathrm{mod}\,2\pi),\,L\big) \;\leq\; \frac{\pi}{4}\left\{ \frac{V_1(\omega_0,\phi_0)}{\min_{x\geq a} g'(x)} + \sup_{x\geq a}\frac{|g''(x)|}{|g'(x)|^2} \right\}.$$

The results obtained for the scalar case in Sect. 5.1 extend to the problem studied in this section in a similar way.

This section concludes with some remarks on Hopf's work on the problem discussed in this section (see Hopf, 1936, p.187ff). He assumed the force only depends on the position: $F(\omega) = \chi(\phi)$, and showed that the wheel's final position converges in the weak-star topology to a distribution with density given by $\chi(\phi)$. However, such a system does not stay at rest once its velocity is zero. In fact it is well known (see, for example, Arnold, 1978 , p.15ff) that it necessarily is conservative. Hopf himself points out that an analogous result may be derived when the force has the form $F(\phi,\omega) = r(\omega)\omega(\phi)$ (see Hopf, 1936, p. 188). That is precisely what has been done in this section.

In his 1937 paper, Hopf gives a fifteen page proof of convergence for what appears to be a much more general class of forces than those considered in this section and shows that the limiting density evaluated at ϕ in $[0,2\pi]$ is proportional to $F(\phi,0)$. Again there are problems from a physical point of view. Either $F(\phi,\omega)$ is discontinuous at $\omega = 0$ or the system does not stay at rest once $\omega = 0$. I have not been able to find a reasonable set of assumptions under which this result remains valid.

5.4 Statistical Regularity of a Dynamical System

Hopf (1934, p.66) introduced the notion of *statistical regularity* to make precise the idea of unpredictability of a conservative dynamical system. In this section these ideas are extended to an arbitrary dynamical system. When the weak-star topology is considered the resulting concept corresponds to that of mixing sequence of random variables introduced by Rényi and Révész (1958).

Two simple yet useful theorems that characterize statistically regular dynamical systems are proved in this section. They are used to show that the dynamical systems studied in the previous sections of this chapter are statistically regular.

Definition. Assume T is a subset of the real numbers and E an open subset of $\mathbb{R}^n$. The family of measurable functions $h(\tau, \cdot) : E \to \mathbb{R}^p$, parameterized by $\tau \in T$, defines a p dimensional *dynamical system*.

Remark. A typical value of a dynamical system is denoted $h(\tau, x)$. The vector x denotes the set of quantities about which there exists uncertainty and that are described using a joint density. In the case of a physical system they usually correspond to initial conditions but might also involve physical constants. The parameter τ is assumed fixed. Among other things it might denote time, a friction coefficient, a location parameter or a scale parameter.

Example 1. Let $h(\mu, \omega_0, \phi_0)$ denote the final position (modulo 2π) of a rotating wheel with initial position ϕ_0 and initial velocity ω_0, that is slowed down by a frictional force $F(\phi, \omega) = \mu r(\omega)\chi(\phi)$ (see the case of weak frictional forces in Sect. 5.3). Then $h(\mu, \omega_0, \phi_0)$ defines a dynamical system.

In this example, the role of τ is played by the friction coefficient μ while (ϕ_0, ω_0) corresponds to x.

Example 2. Let $h(a, \phi_0, \omega_0)$ denote the final position (modulo 2π) of a rotating wheel with initial position ϕ_0 and initial velocity $\omega_0 + a$, slowed down by a frictional force $F(\phi, \omega) = r(\omega)\chi(\phi)$ (see the location case in Sect. 5.3). The dynamical system corresponding to the scalar case is defined analogously.

Remark. As can be seen in the previous two examples, the dimensions of x and $h(\tau, x)$ need not necessarily be the same.

Definition. Let $h(\tau, x)$ be a dynamical system that maps $T \times E$ into $\mathbb{R}^p$, where $T \subset \mathbb{R}$ and E an open subset of $\mathbb{R}^n$. The system is *statistically regular* if there exists a probability P on $\mathbb{R}^n$ such that for every absolutely continuous probability Q on E the image of Q under $h(\tau, \cdot)$ converges weak-star to P as τ tends to τ_0. If convergence is in total variation, the system is *strongly statistically regular*.

Remark. It may happen that P and Q are the marginal distributions of random variables L and X defined on a common probability space. Then one can inquire about other modes of convergence.

Example. Theorem 4.2 implies that the n dimensional dynamical system $h(t, x) = (tx)(\bmod 1)$ is statistically regular as t tends to infinity. The limiting distribution is uniform on $[0, 1]^n$.

Remark 1. It is implicitly assumed that τ tends to τ_0 along elements in T.

Remark 2. If the limiting distribution P in the definition of statistical regularity is not constant and the parameter τ is time, the concept of statistical regularity provides a way of making precise the idea of unpredictability of a dynamical system: no matter what absolutely continuous distribution Q is used to describe initial conditions, as time passes the position of the dynamical system necessarily approaches a given, fixed non degenerate distribution that does not depend on Q.

Remark 3. The definition of statistical regularity can be extended to the case where the image under $h(\tau, \cdot)$ of any distribution Q that is absolutely continuous with respect to a fixed measure F – that is not Lebesgue measure – is considered. For example, in certain probabilistic applications it is useful to take E discrete and let counting measure play the role of Lebesgue measure. This line of thought is not pursued in this chapter (except for remarks like this one) because all applications of the method of arbitrary functions that involve the results to be proved consider the case of Lebesgue measure.

Remark 4. The parameter set T in the definition of a dynamical system could be any metric space. In Sect. 5.5 dynamical systems where T is a subset of $\mathbb{R}^k$ naturally arise when considering the product of dynamical systems. $\square$

The following theorem provides a characterization of statistically regular dynamical systems. It is an elementary result that gives a unified proof for the remaining theorems considered in this chapter.

Theorem 5.1 *Assume E is an open subset of $\mathbb{R}^n$. Let $\mathcal{F}$ denote a set of absolutely continuous distributions on E such that its convex hull is dense (with respect to L^1- distance) in the set of absolutely continuous distributions on E. That is, given any density $f(x)$ on E and $\varepsilon > 0$, there exist $\lambda_1, \ldots, \lambda_m$ in $[0, 1]$ adding to one and densities $f_1(x), \ldots, f_m(x)$ such that*

$$\int_E \left| f(x) - \sum \lambda_i f_i(x) \right| dx \leq \varepsilon.$$

Let $h(\tau, x)$ denote a dynamical system that maps $T \times E$ into $\mathbb{R}^p$. If there exists a distribution P in $\mathbb{R}^p$ such that the image of any probability Q in $\mathcal{F}$ under $h(\tau, \cdot)$ converges weak-star to P as τ tends to τ_0, then $h(\tau, x)$ is statistically regular. If convergence is in the variation distance, $h(\tau, x)$ is strongly statistically regular.

Proof. Both proofs are similar so only the case of variation distance is considered. What has to be shown is that given any absolutely continuous distribution Q on E, the image of Q under $h(\tau, \cdot)$ converges in the variation distance to P.

Assume Q has density $f(x)$ and let $\varepsilon > 0$ be fixed. Denote by $g_\varepsilon(x) = \sum_{i=1}^m \lambda_i g_i(x)$ a convex linear combination of densities in $\mathcal{F}$ whose L^1-distance to $f(x)$ is less than ε. Denote the distribution function of $g_\varepsilon(x)$ by Q_ε and that of $g_i(x)$ by Q_i. Then

$$Q_\varepsilon\{h(\tau, x) \in A\} = \sum \lambda_i Q_i\{h(t, x) \in A\}. \tag{5.41}$$

Therefore

$$|Q\{h(t, x) \in A\} - P\{h(\tau, x) \in A\}|$$
$$\leq |Q\{h(t, x) \in A\} - Q_\varepsilon\{h(t, x) \in A\}| + |Q_\varepsilon\{h(t, x) \in A\} - P\{h(\tau, x) \in A\}|.$$

Using Proposition 2.6 and (5.41) to bound the first and second terms, respectively:

$$\leq 2\varepsilon + \sum_i \lambda_i |Q_i\{h(\tau, x) \in A\} - P\{h(\tau, x) \in A\}|.$$

By the definition of variation distance:

$$\leq 2\varepsilon + \sum \lambda_i d_V(h(\tau, Q_i), P). \tag{5.42}$$

Letting τ tend to τ_0 in (5.42) and using the hypothesis according to which $h(\tau, Q_i)$ converges in the variation distance to P leads to

$$|Q\{h(t, x) \in A\} - P\{h(\tau, x) \in A\}| \leq 2\varepsilon. \tag{5.43}$$

The proof concludes by taking supremum over all measurable sets A in (5.43) and noting that $\varepsilon > 0$ was arbitrary. $\qquad\square$

Remark 1. It should be noted that both when convergence in the weak-star topology and convergence in the variation distance are considered, the theorem requires that the metric with respect to which $\mathcal{F}$ is dense in the set of all densities on E be L^1, or equivalently, the variation distance.

Remark 2. The theorem is still valid if all distributions considered are absolutely continuous with respect to any fixed measure F on E other than Lebesgue measure. The proof is the same. $\qquad\square$

Borel (1909; 1956, p.97ff) showed that given any two dimensional distribution X uniform on a rectangle with sides parallel to the coordinate axis, the distribution of $(tX)(\mathrm{mod}\,1)$ converges in the weak-star topology to a distribution uniform on the unit square as t tends to infinity. He says that assuming X uniform is not essential due to an argument of Poincaré's. Most of Hopf's proofs of statistical regularity are rigorously reduced to the case where the initial density is uniform. These ideas are made precise in the following theorem.

Theorem 5.2 (Poincaré, Borel, Hopf) *Let $h(\tau, x)$ denote a dynamical system that maps $T \times E$ into $\mathbb{R}^p$, where $T \subset \mathbb{R}$ and E is an open subset of $\mathbb{R}^n$. Assume there exists* *a probability P on $\mathbb{R}^p$ such that the image under $h(\tau, \cdot)$ of any distribution uniform on a hypercube included in E with sides parallel to the coordinate axis converges weak-star to P as τ tends to τ_0. Then $h(\tau, x)$ is statistically regular. If convergence is in the variation distance, the system $h(\tau, x)$ is strongly statistically regular.*

Proof. Theorem 5.1 reduces the proof to showing that the convex hull of the set of densities on E that are uniform on hypercubes (with sides parallel to the coordinate axis) is dense (in the L^1-distance) in the set of all densities on E. It is well known that any density may be approximated arbitrarily well (with respect to L^1-distance) by a continuous density with compact support. Therefore it suffices to show that any continuous density supported by $[0,1]^n$ can be approximated by a convex combination of uniform densities. For the sake of clarity this is proved in the case where $n = 2$. The same proof holds in the n dimensional case.

Let $f(x,y)$ be a continuous density defined on $[0,1]^2$. The unit square is partitioned into n^2 squares of area $1/n^2$ and $f(x,y)$ is approximated by a convex linear combination of uniform densities on these squares, with coefficients proportional to the integral of $f(x,y)$ over the corresponding square. That is, let

$$I_{i,j}^n = [\frac{i-1}{n}, \frac{i}{n}] \times [\frac{j-1}{n}, \frac{j}{n}] \; ;$$

$$\lambda_{i,j}^n = \int_{I_{i,j}^n} f(x,y)dxdy \; ;$$

$$f_{i,j}^n(x,y) = n^2 I\{(x,y) \in I_{i,j}^n\},$$

with I denoting an indicator function and $i,j = 1,\ldots,n$. Define

$$f_n(x,y) = \sum_{i,j} \lambda_{i,j}^n f_{i,j}^n(x,y).$$

The function $f_n(x,y)$ is a convex linear combination of densities uniform on squares with their sides parallel to the coordinate axis. Letting

$$I_n(x,y) = ([nx]/n, ([nx]+1)/n)) \times ([ny]/n, ([ny]+1)/n),$$

it follows that

$$f_n(x,y) = n^2 \int_{I_n(x,y)} f(x,y)dxdy.$$

Hence

$$\min_{(x,y)\in I_n(x,y)} f(x,y) \leq f(x,y) \leq \max_{(x,y)\in I_n(x,y)} f(x,y).$$

Continuity of $f(x,y)$ implies that $f_n(x,y)$ converges pointwise to $f(x,y)$. Convergence in the L^1-distance follows from Scheffé's Theorem (see Billingsley, 1986, p.218). □

Remark. The fact that simple functions are dense in the L^1-space associated to any fixed measure F (see Royden, 1968, p.244) implies that Theorem 5.2 may be generalized to the case where distributions that are absolutely continuous with respect to F are considered. Let $\mathcal{B}$ denote a basis of E, that is, every open set in E may be written as a union of sets in $\mathcal{B}$. It then suffices to show that the image under $h(\tau, \cdot)$ of any distribution with Radon-Nykodym derivative (with respect to F) proportional to the characteristic function of a set B in $\mathcal{B}$ converges weak-star (in variation distance) to a probability P to conclude that the image under $h(\tau, \cdot)$ of any distribution absolutely continuous with respect to F converges weak-star (in variation distance).   □

The following theorem states that the random vector $(tX)(\mathrm{mod}\,1)$ converges in the variation distance to a distribution uniform on $[0,1]^n$ if and only if X has a density (with respect to Lebesgue measure). It is an immediate consequence of Theorem 5.2.

Theorem 5.3 *Let X denote an n dimensional random vector and U_n a distribution uniform on $[0,1]^n$. Then $(tX)(\mathrm{mod}\,1)$ converges in the variation distance to U_n if and only if X is absolutely continuous, that is, has a density with respect to Lebesgue measure.*

Proof. Assume X is uniform on a hypercube of $\mathbb{R}^n$. Theorem 4.9 implies that the random vector $(tX)(\mathrm{mod}\,1)$ converges to U_n in the variation distance and by Theorem 5.2 it now follows that $(tX)(\mathrm{mod}\,1)$ converges to U_n for any absolutely continuous random vector X.

To show that $(tX)(\mathrm{mod}\,1)$ cannot converge to U_n if X has a singular component, note that if

$$F(x) = \lambda F_{ac}(x) + (1 - \lambda)F_s(x)$$

is the Lebesgue decomposition of the distribution function $F(x)$ of X into an absolutely continuous and a discrete component (see Billingsley, 1986, p.435), where $\lambda < 1$, there exists a set S of zero Lebesgue measure such that X belongs to this set with positive probability $(1 - \lambda)$. Hence, for any t there exists a set S_t that has zero Lebesgue measure and satisfies

$$\mathrm{Pr}\{tX \in S_t\} = 1 - \lambda. \tag{5.44}$$

Denote the distribution function of $X(\mathrm{mod}\,1)$ by $F_1(x)$. Given a measurable set A in $[0, 2\pi]$:

$$F_1(A) = \sum_{k \in \mathbf{Z}} F(A + k), \tag{5.45}$$

where $A + k$ denotes the translation of the set A by k units. Equations (5.44) and (5.45) and the fact that a countable union of sets of measure zero has measure zero imply that there exists a set S_t^* with zero Lebesgue measure that includes $(tX)(\mathrm{mod}\,1)$ with probability $(1 - \lambda)$. It follows that the variation distance between $(tX)(\mathrm{mod}\,1)$ and U_n is bounded from below by $(1 - \lambda)$ and no convergence in the variation distance may take place if $\lambda > 0$.   □

Applications

1. Spinning wheel: location case. Let $h(a, \phi_0, \omega_0)$ denote the final position (modulo 2π) of a rotating wheel with initial position ϕ_0, initial velocity $\omega_0 > 0$ and slowed down by a frictional force equal to $F(\phi, \omega) = r(\omega)\chi(\phi)$, where $r(\omega)$ and $\chi(\phi)$ are positive and

$\int_0^{2\pi} \chi(\phi) d\phi = 2\pi$. Suppose that, as ω tends to zero, $r(\omega)$ tends to zero at a rate slower than $\omega^{2-\epsilon}$ for some $\epsilon > 0$, so that

$$g(\omega) = \int_0^\omega \frac{\omega}{r(\omega)} d\omega$$

is well defined. Assume that

$$\lim_{\omega \to +\infty} g'(\omega) = +\infty, \tag{5.46}$$

and either

$$\int_M^{+\infty} \frac{|g''(\omega)|}{|g'(\omega)|^2} d\omega < +\infty \quad \text{for some } M > 0 \tag{5.47}$$

or

$$\lim_{a \to \infty} \sup_{x \geq a} \frac{|g''(x)|}{|g'(x)|^2} = 0. \tag{5.48}$$

Let L denote a distribution on $[0, 2\pi]$ with density $\chi(\phi)$. To prove that this dynamical system exhibits strong statistical regularity, it suffices to show that when ω_0 has a uniform density on a positive interval and ϕ_0 is any density independent of ω_0, the system's final position converges to L in the variation distance as a tends to infinity.

From the results in Sects. 5.1 and 5.3 it follows that (see (5.5) and note that $V_1(g(\omega_0), \phi_0)$ is equal to the total variation of the random variable $g(\omega_0)$, $V(g(\omega_0))$, because ω_0 and ϕ_0 are independent):

$$d_V(\phi_1(\mathrm{mod}\, 2\pi), L) \leq \frac{\pi V(g(\omega_0 + a))}{4}.$$

Theorem 5.2 implies that it suffices to show that provided (5.46) and either (5.47) or (5.48) hold, the total variation of $g(U + a)$ tends to zero as a tends to infinity for U denoting a distribution uniform on $[m, M]$, where $0 \leq m < M$ are arbitrary. Proposition 3.8.b) implies that

$$V(g(U + a)) = \frac{1}{M - m} \left(\frac{1}{g'(m + a)} + \frac{1}{g'(M + a)} + \int_{m+a}^{M+a} \frac{|g''(\omega)|}{|g'(\omega)|^2} d\omega \right)$$

and the required conclusion follows.

To see that hypothesis (5.46) is not enough to have statistical regularity and that therefore additional smoothness assumptions like (5.47) or (5.48) are needed, a function $g(x)$ satisfying (5.46) is now constructed, such that if X is uniform on $[0, 1]$, the distribution of $g(X + a)(\mathrm{mod}\, 1)$ does not have a limit (in the weak-star topology, and therefore in the variation distance) as a tends to infinity.

The function $g(x)$ is defined as follows. For $k \leq x \leq k + 1$, k a positive integer, $g(x)$ takes values between $(2^k - 1)$ and $(2^{k+1} - 1)$ and is continuous, piecewise linear on intervals of length $1/2^{k+1}$ with slopes alternatingly equal to $2^{k+1}/3$ and $2^{k+2}/3$ (see Fig. 5.5).

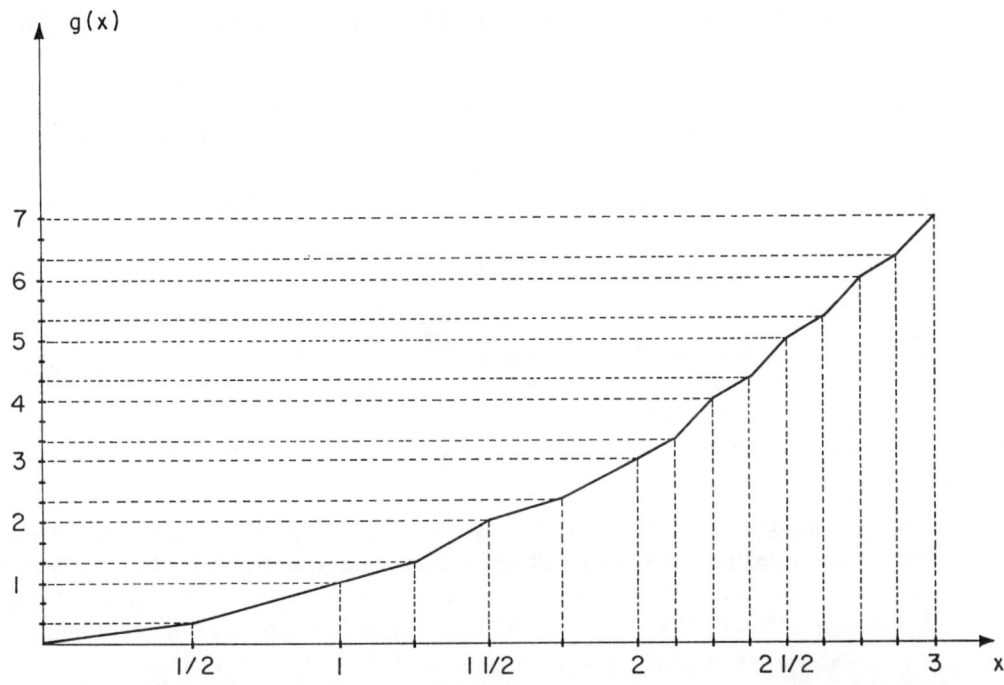

Fig. 5.5. Graph of function leading to the counterexample

Strictly speaking, $g(x)$ is smoothed at the points where its derivative is discontinuous. This can be done in such a way that its first derivative still tends to infinity. Probabilities calculated using the smooth version and the original function differ negligibly. It is clear that the second derivative of the smooth version oscillates wildly.

Let

$$I_{j,k} = \left[k + \frac{j}{2^{k+1}} , \; k + \frac{j+1}{2^{k+1}} \right].$$

Note that

$$g(x)(\mathrm{mod}\,1) \in [0, 1/3] \qquad \text{if and only if} \qquad x \in \bigcup_{k=0}^{+\infty} \bigcup_{j=0}^{2^k - 1} I_{2j,k},$$

and therefore, if ω_0 is uniform on $[0,1]$ and l is a positive integer:

$$\Pr\{g(X+l)(\mathrm{mod}\,1) \in [0, \tfrac{1}{3}]\} = \Pr\{X + l \in \bigcup_{j=0}^{2^a - 1} I_{2j,l}\} = \sum_{j=0}^{2^l - 1} m(I_{2j,l}) = \frac{1}{2},$$

where m denotes Lebesgue measure. A similar calculation shows that if the limit of $\Pr\{g(X+a) \in [0, 1/3]\}$ is calculated for a tending to infinity along the sequence

$\frac{1}{2}, 1\frac{1}{4}, 2\frac{3}{8}, \ldots$ the limit is $\frac{1}{4}$. It follows that $g(X + a)(\bmod 1)$ does not have a limit as a tends to infinity.

2. Spinning wheel: scalar case. Let $h(b, \phi_0, w_0)$ denote the final position (modulo 2π) of a spinning wheel with initial position ϕ_0, initial velocity $bw_0 > 0$ and slowed down by a frictional force equal to $F(\phi, w) = r(w)\chi(\phi)$, where $r(w)$ and $\chi(\phi)$ are positive and $\int_0^{2\pi} \chi(\phi)d\phi = 2\pi$. Define $g(w)$ as in the previous application and assume it satisfies

$$\lim_{b \to +\infty} \min_{bm \leq w \leq bM} bg'(w) = +\infty,$$

and

$$\lim_{b \to +\infty} \sup_{bm \leq w \leq bM} \frac{|g''(w)|}{|g'(w)|^2} = 0.$$

A similar argument to the one given in Application 1 shows that this dynamical system exhibits strong statistical regularity as b tends to infinity. The limiting distribution has density $\chi(\phi)$ on $[0, 2\pi]$.

3. Spinning wheel: weak frictional forces. Let $h(\mu, w_0, \phi_0)$ denote the final position (modulo 2π) of a rotating wheel with initial position ϕ_0 and initial velocity $w_0 > 0$ that is slowed down by a frictional force equal to $F(\phi, w) = \mu r(w)\chi(\phi)$, where $r(w)$ and $\chi(\phi)$ are positive and $\int_0^{2\pi} \chi(\phi)d\phi = 2\pi$. Suppose that, when w tends to zero, $r(w)$ tends to zero at a rate slower than $w^{2-\epsilon}$ for some $\epsilon > 0$, so that

$$g(w) = \int_0^w \frac{w}{r(w)} dw$$

is well defined. The results from Sects. 5.1 and 5.3 imply that

$$\psi_1 = \psi_0 + \frac{1}{\mu} g(w_0),$$

with

$$\psi(\phi) = \int_0^\phi \chi(\phi)d\phi.$$

Proposition 2.6 and the fact that $\phi(\bmod 2\pi)$ can be expressed as a function of $\psi(\bmod 2\pi)$ (see (5.39)) imply that strong statistical regularity follows from showing that for any density w_0 the random variable $\psi_1(\bmod 2\pi)$ converges, in the variation distance, to a distribution uniform on $[0, 1]^n$ as μ tends to zero. Due to Theorem 5.3, it is sufficient to show that the random variable $g(w_0)$ has a density whenever w_0 has one and this follows, without the need of any additional assumption on $g(w_0)$, form Hewitt and Stromberg (1965, Theorem 20.3, p.342).

4. Force as a function of only velocity. Consider a dynamical system whose position at time t, $\phi(t)$, is determined by n coordinates and satisfies the following form of Newton's equation

$$\phi''(t) = \mu F(\phi'(t)),$$

where F is a function of only velocity. Assume the system eventually comes to rest (i.e. is dissipative) and that F is a continuous and positive with $F(0) = 0$. Let $h(\mu, w_0, \phi_0)$ denote the system's final position (modulo 1) when its position and velocity at time zero are ϕ_0 and ω_0, respectively. Essentially the same argument given in the preceding application shows that this dynamical system exhibits strong statistical regularity as the friction coefficient μ tends to zero. The limiting distribution is uniform on the unit hypercube.

5.5 Physical and Statistical Independence

The 1930's were years in which the axiomatization of the calculus of probability played an important role. From a mathematical point of view, what was new with respect to measure theory was the concept of independence. Quoting from Kolmogorov (1933, 1956 p.8ff):

"Historically the independence of experiments and random variables represents the very mathematical concept that has given the theory of probability its peculiar stamp. [...] In consequence, one of the most important problems in the natural sciences is – in addition to the well known one regarding the essence of the concept of probability itself – to make precise the premises which would make it possible to regard any given real events as independent. This question, however, is beyond the scope of this book."

Hopf was very interested in this problem. He mentioned the connection between the method of arbitrary functions and the concepts of physical and statistical independence in every article he wrote on this subject.

Physical independence of two variables was interpreted by Hopf as meaning that their initial distribution is described by a two dimensional density (see Hopf, 1936, p181). Statistical independence of their final values should then be derived. If this program is successful, statistical independence follows from physical independence and need not be assumed.

Consider, for example, the dynamical system studied in Sect. 5.2 of this chapter. Its position at time t, $\phi(t)$, is an n dimensional vector that satisfies Newton's equation

$$\phi''(t) \;=\; \mu F\big(\phi'(t)\big),$$

where the force slowing it down (and eventually bringing it to rest) is a function of only velocity and a friction coefficient μ. It was shown that given any joint density for the system's initial position and velocity, its final position (modulo 1) converges, in the variation distance, to a distribution uniform on $[0,1]^n$ as μ tends to zero. As the coordinates of this limit are independent, no matter how highly correlated the initial values of the system's position and velocity are (as long as they are physically independent), the coordinates describing its final position are approximately independent (in the statistical sense) for small values of the friction coefficient μ. This applies, in particular, to the version of Buffon's needle problem described in Sect. 5.2. The distance between the needle's center of mass and the nearest line to its left and the angle it makes with that line once it stops are (approximately) stochastically independent if frictional forces are sufficiently weak and initial conditions are physically independent, that is, are described by a joint density.

This phenomenon is present in many of the applications considered in Chap. 4. For example, the coordinates describing the position of a heavy symmetric top converge to independent random variables (see Sect. 2.1 of Chap. 4). In this example the limiting distribution is not uniform: the angle describing nutation approaches a distribution that, under suitable conditions, is approximately an arcsine law.

Consider the case where Buffon's needle experiment is performed many, say n times. The distribution of every individual outcome is approximately uniform if the force slowing down the needle only depends on its velocity and its magnitude is small. What can be said about the joint distribution of the n outcomes? If initial conditions for different experiments are stochastically independent, the joint distribution is also approximately uniform. Yet the assumption of independence may be too strong. The person giving the needle its initial impulse might have a tendency to give it a smaller impulse if on the previous trial the needle traveled a large distance and viceversa. In this case, the distributions describing initial conditions for different experiments are not independent. The following theorem shows that if the collection of initial conditions corresponding to all experiments is described by a joint density, the random vector of resulting final positions is approximately uniform. Hopf (1934, p.76) proved it for conservative dynamical systems using the weak-star topology.

Theorem 5.4: Hopf's Independence Theorem *Let $h_i(\tau_i, x_i)$ be a dynamical system that maps $T_i \times E_i$ into $\mathbb{R}^{p_i}$, $i = 1, \ldots, k$. Define the corresponding product system as*

$$h(\tau, x) = \big(h_1(\tau_1, x_1), h_2(\tau_2, x_2), \ldots, h_k(\tau_k, x_k)\big),$$

where $x = (x_1, \ldots, x_k)$ and $\tau = (\tau_1, \ldots, \tau_k)$.

If every $h_i(\tau_i, x_i)$ is statistically regular as τ_i tends to τ_0^i, then their product is statistically regular as τ tends to $\tau_0 = (\tau_0^1, \ldots, \tau_0^k)$. Further, if $h_i(\tau_i, x_i)$ has P_i as limiting distribution, the limiting distribution of $h(\tau, x)$, P, is the product measure of the P_i's. The same result holds when strong statistical regularity is considered.

Proof. Let $n = \sum n_i$ denote the dimension of the set $E = \Pi_{i=1}^k E_i$ where the product system is defined. Due to Theorem 5.2 it suffices to show that the image under $h(\tau, \cdot)$ of any distribution uniform on a hypercube contained in E with sides parallel to the coordinate axis converges to P. This reduces the proof to the case where initial conditions for different systems are independent. For the variation distance the proof then follows from Proposition 2.9.

For the weak-star topology, let $y = (y_1, \ldots, y_k)$ be a vector in $\mathbb{R}^n$ with $y_i \in \mathbb{R}^{n_i}$. Assume Q_i is an absolutely continuous distribution on E_i, $i = 1, \ldots, k$ and denote their product measure by Q. Interpreting inequalities coordinatewise it follows that:

$$|Q\{h(\tau, x) \le y\} - P\{h(\tau, x) \le y\}| =$$
$$= |\Pi_{i=1}^n Q_i\{h_i(\tau_i, x_i) \le y_i\} - \Pi_{i=1}^n P_i\{h_i(\tau_i, x_i) \le y_i\}|.$$

Applying Lemma 1 in Billingsley (1986, p.367):

$$|Q\{h(\tau, x) \leq y\} - P\{h(\tau, x) \leq y\}| \leq$$
$$\leq \sum |Q_i\{h_i(\tau_i, x_i) \leq y_i\} - P_i\{h_i(\tau_i, x_i) \leq x_i\}|.$$

The proof concludes by letting τ tend to τ_0 and using the fact that individual systems are statistically regular. □

Remark 1. The Independence Theorem can be interpreted as saying that physical independence of initial conditions of various statistically regular dynamical systems implies statistical independence of their limiting values.

Remark 2. It should be noted that the proof given above is not generally useful for finding upper bounds for the variation distance (or any other distance metrizing the weak-star topology) between the image under $h(\tau, \cdot)$ of an absolutely continuous distribution Q and the limiting distribution, P, even if upper bounds for the corresponding distance between the individual systems and their limits are available, except in the case where initial conditions are independent. It seems that every product system needs to be studied separately to this effect.

The applications in Chaps. 3 and 4 lead to many examples where product systems arise. They all involve a random vector of the form $(tX)(\mathrm{mod}\,1)$ and possibly some additional random vector Y that does not depend on t. Theorem 4.9 can be used to obtain upper bounds for the variation distance between the product system and its limit if it is possible to apply Proposition 2.8, that is, if the process of conditioning on Y, obtaining upper bounds and then unconditioning via Proposition 2.8, leads to useful bounds. This is often the case: see Sect. 4.2.6 for an example.

Remark 3. Based on the remark following Theorem 5.2 and the fact that the product of a basis for E_1 and a basis for E_2 is a basis for the product space $E_1 \times E_2$, the proof of Theorem 5.4 may be adapted to the general case where the distributions being considered are absolutely continuous with respect to some fixed, arbitrary measure instead of Lebesgue measure.

5.6 The Method of Arbitrary Functions and Ergodic Theory

Hopf not only developed the mathematics of the method of arbitrary functions, but also was one of the pioneering researchers in ergodic theory. In Hopf (1934) he deals with both subjects simultaneously. In this paper (p.70) he introduces the concept of *strong-mixing* for measure-preserving flows and shows that it is equivalent to statistical regularity. This equivalence result is proved in Theorem 5.5.

The possible motions of a conservative dynamical system can always be described in the following way: Let x denote a point in phase space (the coordinates that describe the system's position and velocity at a given instant of time). After time t has elapsed, the point x has moved in phase space to another definite point, $T_t(x)$. The existence of T_t follows from Newton's equations. Trivially, $T_0(x) = x$. As the system is conservative, the forces acting on the system do not depend explicitly on time and therefore $T_t(T_s(x)) = T_{t+s}(x)$. The family of functions $T_t(x)$ defines a dynamical system (in the sense discussed in Sect. 5.4) where the parameter is time and the unknown quantities that are described by a joint density are the initial position and velocity vectors. Note that in this case the space on which the dynamical system is defined is the same as that where it takes its values.

An essential property of conservative mechanisms is that an invariant measure exists in phase space such that the volume included between any two manifolds of constant energy is finite (Liouville's Theorem). This implies the existence of an invariant probability measure on every manifold of constant energy and motivates studying *measure preserving flows*.

Definition. Let $(E, \mathcal{F}, m)$ be a probability space. The family of transformations $\{T_t(x) : E \to E ; t \geq 0\}$ defines a *measure-preserving flow* if

a) $T_0(x) = x$.

b) $T_{t+s}(x) = T_t(T_s(x))$.

c) $m(T_t^{-1}(A)) = m(A)$ for all A in the sigma-algebra $\mathcal{F}$ and for all $t \geq 0$.

Example. Chapters 3 and 4 were dedicated to a detailed study of the random vector $(tX)(\mathrm{mod}\, 1)$ as t tends to infinity. This dynamical system does not define a measure-preserving flow. Yet $(tx)(\mathrm{mod}\, 1)$ is closely related to the second coordinate of the system $T_t(x, y) = (x, (tx + y)(\mathrm{mod}\, 1))$. If this system is viewed as defined on $\mathbb{R}^n \times [0, 1]^n$, it is a measure-preserving flow. Any pair of independent, absolutely continuous distributions (X, Y) define an invariant measure if Y is uniform on $[0, 1]^n$.

Remark 1. The transformations T_t are not required to be invertible. There are many natural measure-preserving flows that are not one-to-one, for example the one dimensional baker's transformation $x \to (2x)(\mathrm{mod}\, 1)$. The set $T_t^{-1}(A)$ that appears in part c) of the definition denotes the preimage of the set A through T_t.

Remark 2. The fact that $T_t(x)$ defines a measure-preserving flow with respect to m means that if the random vector X describing the system's position (in phase space) at time zero has distribution m, the distribution of its position at any time $t, T_t(X)$, is also distributed like m.

Remark 3. In what follows, E is assumed to be an open subset of $\mathbb{R}^n$. The results and proofs are valid in much more general spaces.

Remark 4. All that follows holds for the case where the family of transformations $T^1(x), T^2(x), \ldots$ is considered with $T : E \to E$ measure-preserving and $T^n(x)$ denoting the function that results from composing T with itself n times.   □

A flow $T_t(x)$ that preserves m is said to be *ergodic* if the only sets that remain fixed under T_t have probability zero or one, that is, $T_t^{-1}(A) = A$ implies $m(A) = 0$ or 1. As its name indicates, ergodic theory is devoted to the study of ergodic flows. One of the main notions used in ergodic theory to make precise the idea of unpredictable dynamical system is that of strong-mixing due to Hopf (1934).

Definition. A flow $T_t(x)$ defined on E that preserves m is said to be *strong-mixing* if for every measurable sets A and B included in E

$$\lim_{t \to +\infty} m(T_t^{-1}(A) \cap B) = m(A)m(B). \square$$

Remark. Assume the system's position at time t is described by a random vector X_t with initial distribution equal to m and suppose that $m(B) > 0$. From the fact that

$$m\{T_t(x) \in A | x \in B\} = m(T_t^{-1}(A) \cap B)/m(B)$$

it follows that the flow is strong-mixing if and only if for any measurable set B such that $m(B) > 0$ the distribution of X_t, conditioned on $X_0 \in B$, converges in the weak-star topology to a random vector with the same distribution as X_0. $\square$

The following theorem establishes the equivalence between the concepts of strong-mixing and statistical regularity.

Theorem 5.5 (Hopf) *Assume $T_t(x)$ defines a flow that is measure-preserving with respect to an absolutely continuous probability measure m with support equal to an open subset E of $\mathbb{R}^n$. Then $T_t(x)$ is strong-mixing if and only if it is statistically regular.*

Proof. In the definition of dynamical system take $h(\tau, x) = T_t(x)$ and let X_0 denote a random vector distributed like m.

The fact that $(X_0 | X_0 \in B)$ has a density for any B with $m(B) > 0$ and the remark following the definition of strong-mixing show that statistical regularity implies strong-mixing.

It is now shown that strong-mixing implies statistical regularity. Due to Theorem 5.1 the proof reduces to showing that the set of distributions $\{(X_0 | X_0 \in B) : B$ such that $m(B) > 0\}$ is dense, with respect to L^1-distance, in the set of all densities on E. Denote the density of X_0 by $f(x)$ and let Y be an arbitrary, absolutely continuous distribution on E with density $g(x)$. The hypothesis according to which $f(x)$ has support equal to E implies that Y is absolutely continuous with respect to X_0 and therefore the Radon-Nykodym derivative of Y with respect to X_0, $h(x)$, is well defined (it is equal to $g(x)/f(x)$ except on a set of Lebesgue measure zero). As the density of $(X_0 | X_0 \in B)$ is proportional to $f(x)$ if $x \in B$ and 0 otherwise, all that needs to be shown is that any function $h(x)$ in the L^1-space generated by $f(x)$ (i.e. such that $\int_E |h(x)| f(x) dx < +\infty$) can be approximated arbitrarily well by a linear combination of indicator functions of measurable sets in E and this follows from Royden (1968, p244). $\square$

Remark 1. The proof of Theorem 5.5 is valid in the general case where the distributions considered are absolutely continuous with respect to any fixed measure on a general space (instead of Lebesgue measure).

Remark 2. A flow $T_t(x)$ defined on E that preserves m is said to be *weak-mixing* if for all measurable sets A and B included in E

$$\lim_{t \to +\infty} \frac{1}{t} \int_0^t |m(T_t^{-1}(A) \cap B) - m(A)m(B)| dt = 0.$$

A well known result from ergodic theory (see Walters, 1982, p.45 for the discrete version) states that if E is an abstract measure space with a countable basis (which is the case when E is included in $\mathbb{R}^n$), the flow $T_t(x)$ is weak-mixing if and only if there exists a subset D of the positive real numbers satisfying

$$\lim_{a \to +\infty} \frac{m_L(D \cap [0, a])}{a} = 1, \tag{5.49}$$

where m_L denotes Lebesgue measure, and such that for any measurable sets A and B:

$$\lim_{t \to +\infty} m(T_t^{-1}(A) \cap B) = m(A)m(B),$$

as long as t tends to infinity along a sequence of elements in D. Therefore, from the point of view of unpredictability of a dynamical system, the difference between weak and strong-mixing is of no importance. This result, combined with Theorem 5.5, implies that the flow $T_t(x)$ is weak-mixing if and only if there exists a set D satisfying (5.49) such that the dynamical system $T_t(x)$, $t \in D$, is statistically regular.

Remark 3. The hypothesis that m is invariant was not used in the proof of Theorem 5..5. All that was used was that there exists a random vector X_0 with support equal to E such that $(T_t(X_0)|X_0 \in B)$ converges in the weak-star topology to a distribution m that does not depend on B, for any B such that $m(B) > 0$. Further, the hypothesis that $T_t(x)$ takes values in the same space as x does was not relevant either. The definition of strong-mixing can therefore be extended as follows.

Let $h(\tau, x)$ be a dynamical system that maps $T \times E$ into $\mathbb{R}^p$. The system is *strong-mixing* if there exists an absolutely continuous probability measure m with support equal to E such that for any B with $m(B) > 0$ the image through $h(\tau, \cdot)$ of m conditional on B converges weak-star to a distribution P that does not depend on B. The same notion may also be defined with respect to the variation distance.

The argument given in the proof of Theorem 5.5 can be used to show that a dynamical system is strong-mixing if and only if it is statistically regular. This result is easily extended to distributions that are absolutely continuous with respect to an arbitrary, fixed measure that is not Lebesgue measure.

Remark 4. The notion of mixing sequence of random variables defined by Rényi and Révész (1958) is equivalent to the general concept of strong-mixing introduced in the previous remark. The generalized version of Theorem 5.5 discussed in Remark 3 corresponds to Theorem 1 in Rényi (1958). Rényi and Révész (1958) also found sufficient conditions for a Markov chain to be mixing.   □

This section concludes with an application discussed by Hopf (1934, p.87ff).

Random Rate Systems. Consider a particle undergoing uniform rectilinear motion in a closed vessel C with perfectly reflecting walls. This means that if $T_t(x)$ denotes the particle's position at time t when its initial velocity is $v = 1$, then $T_{vt}(x)$ denotes its position at time t when its velocity is v. Such a flow is invariant and ergodic with respect to a distribution uniform on C. If the particle's initial velocity is known and only its initial position is assumed random, the distribution of its position does not necessarily converge to a distribution uniform on C as time passes (see e.g. the billiard example discussed in Sect. 4.2.3). Hopf (1934) showed that if the particle's initial velocity also is assumed random and described by a density, then the distribution of the particle's position does converge to a distribution uniform on C as t tends to infinity. The following theorem makes this precise:

Theorem 5.6 *Let $T_t(x)$ define an ergodic flow on the open set E included in $\mathbb{R}^n$ that preserves the absolutely continuous probability measure m. Then the dynamical system $h(t, x, v) = T_{vt}(x)$ is statistically regular, as t tends to infinity. That is, an ergodic flow evolving at a random rate is strong-mixing. The limiting distribution is equal to m.*

Proof. Due to Theorem 5.2 it suffices to show that if X has a uniform distribution on $[x_0, x_1]$ and V a uniform distribution on $[v_0, v_1]$ independent of X, then for any measurable set A in E the probability that the random vector $T_{Vt}(X)$ belongs to A tends to $m(A)$ as t tends to infinity.

As V is independent of X and uniform

$$\Pr\{T_{Vt}(X) \in A\} = c_1 \int_{v_0}^{v_1} \Pr\{T_{vt}(X)\} dx,$$

where $c_1 = 1/(v_1 - v_0)$. And as X is uniform:

$$\Pr\{T_{Vt}(X) \in A\} = c_1 c_2 \int_{v_0}^{v_1} \int_{x_0}^{x_1} I\{T_{vt}(x) \in A\} dx,$$

where $c_2 = 1/(x_1 - x_0)$ and I denotes an indicator function. Applying Tonelli's Theorem (see Billingsley, 1986, p.238) and making a change of variable:

$$\Pr\{T_{Vt}(X) \in A\} = c_1 c_2 \int_{x_0}^{x_1} \left(\frac{1}{t} \int_{v_0 t}^{v_1 t} I\{T_u(x) \in A\} du \right) dx. \tag{5.50}$$

Due to Birkhoff's Ergodic Theorem,

$$\lim_{t \to +\infty} \frac{1}{t} \int_0^t I\{T_u(x) \in A\} du = m(A). \tag{5.51}$$

Letting t tend to infinity in (5.50) and applying (5.51) and Lebesgue's Dominated Convergence Theorem (the dominating function being constant and equal to $(v_1 - v_0)$) concludes the proof.   □

5.7 Partial Statistical Regularity

Consider, once again, a simple harmonic oscillator. A block is attached to an ideal spring and free to move on a frictionless horizontal table. The spring is stretched a distance A from its equilibrium position and released from rest (see Chap. 1). If the oscillator's amplitude and angular frequency are described by random variables A and ω having a joint density, it was proved that the spring's displacement converges to a scale mixture of arcsine laws, the mixing distribution being equal to A. As the limiting random variable does involve A, this dynamical system is not statistically regular. Yet dependence on the spring's angular frequency has washed away. Further, the limit does exhibit some kind of statistical regularity because the scale mixture determined by A always involves an arcsine law. For another examples exhibiting the same type of behavior, see the general case of small oscillations discussed in Sect. 4.2.2.

In this section, the concept of statistical regularity is extended to cover the examples discussed above. It is shown that Theorems 5.1, 5.2 and 5.4 apply to the generalized concept, that is, that they remain valid if "statistical regularity" is replaced by "partial statistical regularity." It is suggested that Sect. 5.4 be read before this section, as it is not self contained.

Definition. Let $h(\tau, x)$ be a dynamical system that maps $T \times E$ into $\mathbb{R}^p$ where $T \subset \mathbb{R}$ and E is an open subset of $\mathbb{R}^n$. The system is *partially statistically regular* if there exists a probability P on $\mathbb{R}^q$ and a measurable function g from $\mathbb{R}^q \times E$ to $\mathbb{R}^p$ such that for every absolutely continuous probability Q on E the image of Q under $h(\tau, \cdot)$ converges weak-star to the image of the product measure of P and Q under g as τ tends to τ_0. If convergence is in total variation, the system is said to be *partially strongly statistically regular.*

Example. Consider the dynamical system whose position at time t is given by $x(t) = A\cos(\omega t)$, that is, the simple harmonic oscillator studied in the Sect. 1.1. In this case τ denotes time and x is equal to (A, ω). The sets T and E are $[0, +\infty)$ and $[0, +\infty)^2$, respectively. The integers p and q are both equal to one. As $x(t)$ converges to the product of the random variable describing uncertainty about the initial amplitude and an independent arcsine law S, S plays the role of P and $g(P, (Q_1, Q_2)) = PQ_1$, where PQ_1 has the same distribution as the product of independent random variables distributed like P and Q_1. It follows that this system is partially statistically regular even though it is not statistically regular.

Remark 1. Suppose Q has density $f(x)$. Then the product measure $P \times Q$ and the function g induce the following law on $\mathbb{R}^p$:

$$(P \times Q)\{(l, x) : g(l, x) \in A\} = \int P\{l : g(l, x) \in A\} f(x) dx.$$

Therefore the limiting distribution of the dynamical system is a mixture of the parametric family $\{g(P, x) : x \in E\}$ with mixing density $f(x)$. This covers the case of small oscillations mentioned above: $g(A, \omega, x) = Ax$.

Remark 2. Assume a given dynamical system is partially statistically regular as time tends to infinity, P is not a degenerate distribution, and $g(P, Q)$ does depend on the distribution of P. The concept of partial statistical regularity may then be interpreted as meaning that the system's position becomes to some extent unpredictable as time passes: the distribution it approaches not only depends on the density describing initial conditions but also on a distribution that is independent of initial conditions and therefore cannot be predicted.

Remark 3. If the limiting distribution $g(P, Q)$ does not involve a given coordinate of Q, this may be interpreted as meaning that dependence on that coordinate washes away as τ tends to τ_0. For example, dependence on the harmonic oscillator's angular frequency washes away as time passes.

That a dynamical system may exhibit partial statistical regularity in the non trivial sense described in the preceding remark even if dependence on none of its coordinates washes away, can be seen from considering the bouncing ball example studied in Sect. 2.1 of Chap. 3.   □

All characterization theorems proved for statistically regular systems can be extended to systems that exhibit partial statistical regularity. This section concludes with some remarks regarding these extensions.

Remark 1. The proof of the generalized version of Theorem 5.1 (see Sect. 5.4) requires some extra refinements. First consider the variation distance. With the same notation introduced in the proof of Theorem 5.1 and the obvious notation for product measures:

$$|Q\{h(\tau, x) \in A\} - (P \times Q)\{g(l, x) \in A\}| \leq$$
$$\leq |Q\{h(\tau, x) \in A\} - Q_\epsilon\{h(\tau, x) \in A\}| +$$
$$|Q_\epsilon\{h(\tau, x) \in A\} - (P \times Q_\epsilon)\{g(l, x) \in A\}| +$$
$$|(P \times Q_\epsilon)\{g(l, x) \in A\} - (P \times Q)\{g(l, x) \in A\}|.$$

The first term is less than 2ε due to the definition of Q_ϵ. To bound the second term, condition on P, apply the same argument used to bound the second term in the original proof and then uncondition. The fact that Q_ϵ may be chosen independent of P is important for this step. The third term also has 2ε as an upper bound. To see this, condition on P, apply Proposition 2.6 and then uncondition. The fact that P and Q are independent has been used in this final step. The rest of the proof proceeds as in Theorem 5.1.

In the case of the weak-star topology care must be taken to ensure that, given an absolutely continuous distribution Q on E, the continuity sets of $g(P, Q)$ are included in those of $g(P,Y)$ for any Y in $\mathcal{F}$. If $g(P,Q)$ is absolutely continuous, this is trivially the case.

It should be noted that $g(P,Q)$ can be absolutely continuous even if P is not. For example, if $g(p,Q)$ is absolutely continuous, with density $f_p(x)$, for p belonging to a set that to which P assigns probability one, a simple calculation based on the independence of Q and P and Tonelli's Theorem (Billingsley, 1986, p.238) imply that $g(P,Q)$ is absolutely continuous, with density evaluated at x equal to $\int f_p(x)dF_P(p)$, where F_P denotes the distribution function of P.

Remark 2. Theorems 5.2 and 5.5 (see Sects. 5.4 and 5.5) also remain valid if "statistical regularity" is replaced by "partial statistical regularity." The corresponding proofs do not require any modification.

6. Non Diagonal Case

Chapter 4 provided a detailed study of the behavior of the random vector $(tX)(\operatorname{mod} 1)$ for large values of t. From the fact that

$$
tX = \begin{pmatrix} tX_1 \\ tX_2 \\ \vdots \\ tX_n \end{pmatrix} = \begin{pmatrix} t & 0 & \cdots & 0 \\ 0 & t & \cdots & 0 \\ \vdots & \vdots & \ddots & \vdots \\ 0 & 0 & \cdots & t \end{pmatrix} \begin{pmatrix} X_1 \\ X_2 \\ \vdots \\ X_n \end{pmatrix},
$$

this may be viewed as the *"diagonal case."* This chapter deals with the general case, that is, given an n dimensional random vector X and a collection of n by n matrices, $\{A(\tau);\ \tau \in T \subset \mathbb{R}\}$, attention is focused on conditions under which $(A(\tau)X)(\operatorname{mod} 1)$ converges to a distribution uniform on the unit hypercube, U_n, as τ tends to infinity.

If the absolute value of all eigenvalues of the matrices $A(\tau)$ tend to infinity as τ does, bounds on the variation distance between $(A(\tau)X)(\operatorname{mod} 1)$ and U_n that only depend on the norm of the inverse of $A(\tau)$ and a measure of smoothness of the density of X are derived in Theorem 6.2. Section 6.1.1 is dedicated to this result and its consequences.

The case where the absolute value of some eigenvalues tends to infinity while others remain bounded or tend to zero is considered in Sect. 6.1.2. Special attention is given to two particular situations: $(A^k X)(\operatorname{mod} 1)$ and $(e^{tA} X)(\operatorname{mod} 1)$. In Theorems 6.10 and 6.14, necessary and sufficient conditions for weak-star convergence of $(A^k X)(\operatorname{mod} 1)$ and $(e^{tA} X)(\operatorname{mod} 1)$ to a distribution uniform on $[0,1]^n$ as the integer k and real number t tend to infinity, respectively, are derived. They involve number theoretic properties of the generalized eigenvectors of the matrix A. Proposition 6.16 provides upper bounds for the corresponding rates of convergence in the two dimensional case.

The theory developed in Sect. 6.1 is applied to systems of linear differential equations (Sect. 6.2) and automorphisms of the n dimensional torus (Sect. 6.3).

6.1 Mathematical Results

6.1.1 Convergence in the Variation Distance

Notation   Given a random vector $X = (X_1, \ldots, X_n)'$, the total variation of X_i conditioned on the random vector composed of the remaining $(n-1)$ components of X

is noted $V_i(X_1, \ldots, X_n)$ or simply $V_i(X)$. That is, the subindex i in $V_i(X)$ indicates that the total variation of the ith component of X is considered, conditioned on the random vector composed of the remaining $(n-1)$ random variables.

Definition. Given an n by n real valued matrix $B = (b_{ij})_{i,j=1}^{n}$, where b_{ij} denotes the jth entry of the ith row, define

$$\|B\|_1 = \max_{1 \leq j \leq n} \sum_{i=1}^{n} |b_{ij}|,$$

$$\|B\|_\infty = \max_{1 \leq i \leq n} \sum_{j=1}^{n} |b_{ij}|.$$

That is, $\|B\|_1$ and $\|B\|_\infty$ are the largest among the L^1 norms of the columns and rows of B, respectively.

Lemma 6.1 *Assume the random vector* $X = (X_1, \ldots, X_n)'$ *has a density,* $f(x)$, *with continuous, integrable, first order partial derivatives. Let A be an invertible matrix with inverse $A^{-1} = (a^{ij})_{i,j=1}^{n}$ and r_i and c_j the L^1 norms of the ith row and j^{th} column, respectively, that is*

$$r_i = \sum_{j=1}^{n} |a^{ij}| \, ; \qquad c_j = \sum_{i=1}^{n} |a^{ij}|.$$

Also denote

$$W(X) = \sum_{i=1}^{n} V_i(X) \, ; \qquad M(X) = \max_{1 \leq i \leq n} V_i(X).$$

Then:

a) $\quad V_j(AX) = \displaystyle\int_{\mathbb{R}^n} \left| \sum_{i=1}^{n} a^{ij} \frac{\partial f(x)}{\partial x_i} \right| dx.$

b) $\quad V_j(AX) \leq M(X)c_j \leq M(X)\|A^{-1}\|_1.$

c) $\quad \displaystyle\sum_{j=1}^{n} V_j(AX) \leq \sum_{i=1}^{n} r_i V_i(X) \leq W(X)\|A^{-1}\|_\infty.$

Proof. From Proposition 4.5.d):

$$V_j(AX) = \int \left| \frac{\partial g(x)}{\partial x_j} \right| dx, \qquad (6.1)$$

where $g(x)$ denotes the density of AX. Using the change of variable formula and the chain rule in (6.1) proves a). Formulas b) and c) then follow easily. $\quad\square$

Theorem 6.2 *Under the assumptions and notation introduced in the preceding Lemma:*

a) $d_V((AX)(\mathrm{mod}\,1),\,U_n) \;\leq\; \frac{1}{8}M(X)\sum_{j=1}^{n} c_j \;\leq\; \frac{n}{8}M(X)\|A^{-1}\|_1.$

b) $d_V((AX)(\mathrm{mod}\,1),\,U_n) \;\leq\; \frac{1}{8}\sum_{i=1}^{n} r_i V_i(X) \;\leq\; \frac{1}{8}W(X)\|A^{-1}\|_\infty.$

Proof. From Theorem 4.9 and Proposition 4.6:

$$d_V((AX)(\mathrm{mod}\,1),\,U_n) \;\leq\; \frac{1}{8}\sum_{j=1}^{n} V_j(AX),$$

and the proof now follows from Lemma 6.1.   □

Corollary. *Assume $\{A(\tau);\ \tau \in T \subset \mathbb{R}\}$ is a collection of invertible, n by n matrices and X an n dimensional random vector satisfying the hypothesis of Lemma 6.1. Then:*

a) $d_V((A(\tau)X)(\mathrm{mod}\,1),\,U_n) \;\leq\; \frac{n}{8}M(X)\|A(\tau)^{-1}\|_1.$

b) $d_V((A(\tau)X)(\mathrm{mod}\,1),\,U_n) \;\leq\; \frac{1}{8}W(X)\|A(\tau)^{-1}\|_\infty.$

Proof. Follows directly from Theorem 6.2.   □

Remark. It follows from the Corollary that a sufficient condition for the random vector $(A(\tau)X)(\mathrm{mod}\,1)$ to converge to U_n as τ tends to infinity is that $A(\tau)^{-1}$ tend to the zero matrix. In particular, a sufficient condition for convergence of $(A^k X)(\mathrm{mod}\,1)$ to U_n in the variation distance is that all the eigenvalues of A lie outside the unit circle. Similarly, $(e^{tA}X)(\mathrm{mod}\,1)$ converges to U_n in the weak-star topology if the real part of all eigenvalues of A is positive.   □

The following result generalizes Theorem 6.2 to the case where A is not an invertible square matrix but has more columns than rows and is of full rank.

Theorem 6.3 *Assume A is a full rank m by n matrix with $m < n$ and X an n dimensional random vector whose density has continuous, integrable, first order partial derivatives.*

As the matrix A is of full rank, it has m linearly independent columns. Assume, without loss of generality, that they are the first m columns, and denote the submatrix they form by B. Appropriate measures of smoothness are then given by

$$M(X) = \max_{1 \leq i \leq m} V_i(X)\,; \qquad W(X) = \sum_{i=1}^{m} V_i(X).$$

Then:

a) $d_V((AX)(\mathrm{mod}\,1),\,U_m) \;\leq\; \frac{m}{8}M(X)\|B^{-1}\|_1.$

b) $d_V((AX)(\mathrm{mod}\,1),\,U_m) \;\leq\; \frac{1}{8}W(X)\|B^{-1}\|_\infty.$

Proof. Let $A = (B\,|\,C)$ and given $\alpha > 0$ define the n by n matrix

$$E_\alpha = \begin{pmatrix} B & C \\ 0 & \alpha I \end{pmatrix},$$

with 0 and I denoting the corresponding null and identity matrices. Elementary linear algebra shows that E_α is invertible and

$$E_\alpha^{-1} = \begin{pmatrix} B^{-1} & -\alpha^{-1}B^{-1}C \\ 0 & \alpha^{-1}I. \end{pmatrix}. \tag{6.2}$$

To prove a) use Proposition 2.6, Theorem 4.9, Proposition 4.6 and Lemma 6.1.a):

$$d_V\big((AX)(\mathrm{mod}\,1),\, U_m\big) \leq d_V\big((E_\alpha X)(\mathrm{mod}\,1),\, U_n\big)$$

$$\leq \frac{1}{8}\sum_{j=1}^{n} V_j(E_\alpha X)$$

$$\leq \frac{1}{8}\sum_{i=1}^{n}\sum_{j=1}^{n} |e_\alpha^{ij}| V_j(X), \tag{6.3}$$

where e_α^{ij} denotes the (i,j)-th element of E_α^{-1}. Denote the (i,j)-th element of B^{-1} by b^{ij}. Letting α tend to infinity in (6.3) and using (6.2) implies that

$$d_V\big((AX)(\mathrm{mod}\,1),\, U_m\big) \leq \frac{1}{8}\sum_{i=1}^{m}\sum_{j=1}^{m} |b^{ij}| V_i(X).$$

The rest of the proof follows the same lines as that of parts b) and c) of Lemma 6.1.   □

Remark. Any submatrix of A consisting of m linearly independent columns can play the role of B in the preceding proof. Therefore

$$d_V\big((AX)(\mathrm{mod}\,1),\, U_m\big) \leq \frac{m}{8} M(X)\min \|B^{-1}\|_1,$$

where the minimum is taken over all such matrices.   □

Let v be an n dimensional vector and X an n dimensional random vector, and denote their inner product by $v\cdot X$. Theorem 6.3 provides an upper bound for the variation distance between $(v\cdot X)(\mathrm{mod}\,1)$ and a distribution uniform on the unit interval. A proof based on results from Chaps. 3 and 4 derives the same bounds under more general hypothesis on X and is therefore preferred.

Proposition 6.4 *Let $X = (X_1,\ldots,X_n)'$ denote an n dimensional random vector with every X_i having bounded variation when conditioned on the vector composed of the remaining $(n-1)$ coordinates of X. Assume $v = (v_1,\ldots,v_n)'$ is a fixed vector in $\mathbb{R}^n$ different from zero and denote the inner product of v and X by $v\cdot X$. Then:*

$$d_V\big((v\cdot X)(\mathrm{mod}\,1),\, U\big) \leq \frac{1}{8}\min_{1\leq i\leq n}\frac{V_i(X)}{|v_i|}.$$

Proof. Theorem 3.9 and Proposition 4.6 imply that

$$
\begin{aligned}
d_V\big((v{\cdot}X)(\mathrm{mod}\,1)\,,\,U\big) \;&\leq\; \frac{1}{8}V(v\cdot X)\\[4pt]
&\leq\; \frac{1}{8}V_1(v\cdot X\,|\,X_2,\ldots,X_n)\\[4pt]
&=\; \frac{1}{8}V_1(v_1X_1,X_2,\ldots,X_n)\\[4pt]
&=\; \frac{1}{8}\frac{V_1(X)}{|v_1|},
\end{aligned}
$$

and the proof now follows from the symmetric role played by the X_i's. $\square$

6.1.2 Weak-star Convergence

This section begins by establishing a general result which provides necessary and sufficient conditions on the matrices $A(\tau)$ for weak-star convergence of $\big(A(\tau)X\big)(\mathrm{mod}\,1)$ to U_n.

Proposition 6.5 *Let $\{A(\tau); \tau \in T\}$ denote a collection of n by n matrices indexed by a subset T of the real numbers. Assume X is an n dimensional random vector whose characteristic function, $\widehat{f}(\lambda)$, vanishes at infinity, that is, $\lim_{\|\lambda\|\to\infty}\widehat{f}(\lambda)=0$. Let U_n denote a distribution uniform on $[0,1]^n$. Then:*

a)   *A sufficient condition for weak-star convergence of $\big(A(\tau)X\big)(\mathrm{mod}\,1)$ to U_n when τ tends to infinity is that, for every $m \in \mathbb{Z}_*^n$, the norm of $A(\tau)'m$ tend to infinity when τ does, where $A(\tau)'$ denotes the transpose of $A(\tau)$ and $\mathbb{Z}_*^n$ the set of all n dimensional integer valued vectors with at least one coordinate different from zero.*

b)   *If the characteristic function of X does not vanish at any point or if the limit does not depend on the distribution of X, the condition established in a) is also necessary.*

Proof. Lemma 4.1 implies that the Fourier coefficient associated to $m \in \mathbb{Z}^n$ is equal to $\widehat{f}(2\pi A(\tau)'m)$. Statements a) and b) now follow from Proposition 2.1. $\square$

Additional assumptions on the form of the matrices $A(\tau)$ are needed to obtain more specific characterizations of weak-star convergence of the random vector $\big(A(\tau)X\big)(\mathrm{mod}\,1)$ to a distribution uniform on $[0,1]^n$. To gain some geometric intuition into this problem, assume the matrices $A(\tau)$ are diagonalizable, with eigenvalues $\lambda_1(\tau),\ldots,\lambda_n(\tau)$, and corresponding eigenvectors $v_1(\tau),\ldots,v_n(\tau)$. It seems intuitively natural that a necessary condition for weak-star convergence of $\big(A(\tau)X\big)(\mathrm{mod}\,1)$ to a distribution uniform on $[0,1]^n$, U_n, is that at least one eigenvalue of $A(\tau)$ tend to plus or minus infinity as τ does. As the eigenvectors of $A(\tau)$ are allowed to depend on τ, this is not necessarily the case, as can be seen by considering

$$
A(\tau)=\begin{pmatrix} \lambda\tau+(1-\tau)\mu & (\mu-\lambda)(\tau-1)/u \\ (\lambda-\mu)\tau u & \lambda(1-\tau)+\tau\mu, \end{pmatrix},
$$

where λ and μ are two different real numbers and u is irrational. Given an integer valued vector $m = (m_1, m_2)'$, the first and second coordinates of $A'(\tau)m$ are equal to $m_1\mu + (m_1 + m_2 u)(\lambda - \mu)\tau$ and $m_2\mu + (m_1 + m_2 u)(\mu - \lambda)(\tau - 1)/u$ respectively, where $A'(\tau)$ denotes the transpose of $A(\tau)$. Proposition 2.1 then implies that $(A(\tau)X)(\text{mod }1)$ converges to a distribution uniform on the unit square for any random vector X having a characteristic function that vanishes at infinity (in particular, having a density). Yet, the eigenvalues of $A(\tau)$ are λ and μ and therefore not only remain bounded but do not depend on τ. What happens is that the corresponding eigenvectors, $(1, u)$ and $(1, u\tau/(\tau - 1))$, are approaching each other as τ tends to infinity. In fact, as τ grows, $(A(\tau)X)(\text{mod }1)$ becomes approximately equal to $(\tau\tilde{X}, u\tau\tilde{X})(\text{mod }1)$, with $\tilde{X} = (\lambda - \mu)(X + Y/u)$, and this random vector converges to U_2 as τ tends to infinity. Fortunately, the applications of the method of arbitrary functions in the non diagonal case either consider $(A^k X)(\text{mod }1)$ for a large integers k or $(e^{tA}X)(\text{mod }1)$ for large real numbers t; in both cases the corresponding eigenvectors do not depend on the parameter tending to infinity.

Assume X is a random vector with a characteristic function vanishing at infinity. The two main results of this section are Theorems 6.10 and 6.14, that provide necessary and sufficient conditions on the matrix A for weak-star convergence of $(A^k X)(\text{mod }1)$ and $(e^{tA}X)(\text{mod }1)$ to a distribution that does not depend on X.

Before deriving the general results, consider a two by two diagonalizable matrix, A, with eigenvalues λ and μ and corresponding eigenvectors $(1, u)$ and $(s, 1)$. Let $m = (m_1, m_2)'$ be an integer valued vector with at least one component different from zero. Then:

$$A'^k m = \frac{1}{1 - us} \begin{pmatrix} (m_1 + m_2 u)\lambda^k - u(m_1 s + m_2)\mu^k \\ -s(m_1 + m_2 u)\lambda^k + (m_1 s + m_2)\mu^k \end{pmatrix}.$$

Theorem 6.2 implies that $(A^k X)(\text{mod }1)$ converges to U_2 if both eigenvalues have absolute value larger than one. If both eigenvalues have absolute values smaller than one, the matrix A^k tends to the zero matrix and therefore $(A^k X)(\text{mod }1)$ converges almost surely to a distribution with all its mass at the origin. What happens if one eigenvalue, say λ, has absolute value larger than one while the other one, μ, has absolute value less than one? If the quantity u, that is the tangent of the angle made by the eigenvector associated to λ with the x axis, is irrational, the absolute value of the first coordinate of $A'^k m$ tends to infinity and therefore $(A^k X)(\text{mod }1)$ converges to U_2. On the other hand, if $u = p/q$ is rational, $\|A'^k m\|$ tends to infinity or zero depending on whether $qm_1 + pm_2$ is different or equal to zero. Hence $(A^k X)(\text{mod }1)$ converges to a distribution with Fourier coefficients associated to $m = (m_1, m_2)'$ equal to one if $qm_1 + pm_2 = 0$ and zero otherwise. This distribution is uniform on the subset of $[0, 1]^2$ determined by $(q, p)(\text{mod }1)$. Figure 6.1 shows this set for $u = 2/3$.

The study of the general case begins by defining the class of distributions to which $(A^k X)(\text{mod }1)$ might converge. These depend on subgroups of $\mathbb{Z}^n$. Let H be a subgroup of $\mathbb{Z}^n$. Then it is well known that there exist $k \leq n$ linearly independent vectors in $\mathbb{Z}^n$, $m_1, \ldots, m_k$, that generate H. That is, every member of H can be written as a unique linear combination of $m_1, \ldots, m_k$ with integer coefficients. Conversely, every such set of generators generates a subgroup of $\mathbb{Z}^n$.

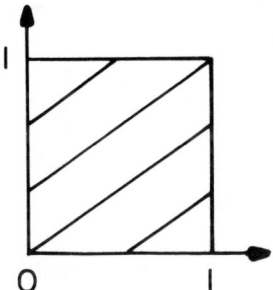

Fig. 6.1. Drawing with parallel lines in unit square

Example. In $\mathbb{Z}^2$, with $k = 1$ and $m_1 = (1, 2)$, the corresponding group is

$$H = \{(x, y) \in \mathbb{Z}^2 : y = 2x\}.$$

Definition. Let $U_1, \ldots, U_k$ be independent, identically distributed random variables uniform on $[0, 1]$. Let $m_1, \ldots, m_k$ denote k vectors in $\mathbb{Z}^n$. The distribution of the random vector $U_H = \left(\sum_{i=1}^{k} U_i m_i \right) (\mathrm{mod}\, 1)$ is called the uniform distribution induced by the group H.

Example. The uniform distribution induced by the group generated by $(1, 2)$ is a distribution uniform on the subset of the unit square shown in Fig. 6.2.

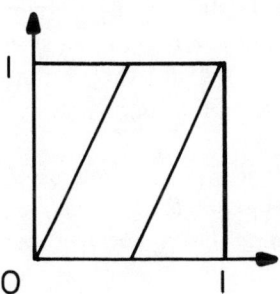

Fig. 6.2. Uniform distribution induced by H

In Proposition 6.6 the Fourier coefficients of U_H are derived and used to show that the definition of U_H is unambiguous, that is, does not depend on the set generating H.

Definition. Let H be a subset of $\mathbb{R}^n$. The set of points in $\mathbb{Z}^n$ orthogonal to H is a subgroup of $\mathbb{Z}^n$, called subgroup orthogonal to H, and denoted by $H^\perp$.

Remark. It is well known that if H itself is a subgroup of $\mathbb{Z}^n$ then $(H^\perp)^\perp$ is equal to H.

Proposition 6.6 *Let H be a subgroup of $\mathbb{Z}^n$ generated by k linearly independent vectors $m_1, \ldots, m_k$. Let $s_1, \ldots, s_r$, $r \geq k$, denote a collection of vectors in $\mathbb{Z}^n$ generating H and $U_1, \ldots, U_r$ be independent, identically distributed random variables uniform on the unit interval. Then:*

a)   *The random vector $\left(\sum_{i=1}^k U_i m_i\right) (\mathrm{mod}\, 1)$ has the same distribution as the random vector $\left(\sum_{i=1}^r U_i s_i\right) (\mathrm{mod}\, 1)$. The definition of U_H is therefore unambiguous.*

  Let $\widehat{f}(\lambda)$ denote the characteristic function of U_H. Then:

b)   *The Fourier coefficients of U_H are equal to*

$$\widehat{f}(2\pi m) = \begin{cases} 1, & m \in H^\perp, \\ 0, & m \notin H^\perp. \end{cases} \tag{6.4}$$

Proof. From Lemma 4.1 and the independence of the U_i's it follows that the Fourier coefficient of $\left(\sum U_i s_i\right)(\mathrm{mod}\, 1)$ evaluated at $l \in \mathbb{Z}^n$ is equal to $\Pi_i \widehat{f}_U(2\pi l \cdot m_i)$, where $\widehat{f}_U(\lambda)$ denotes the characteristic function of a distribution uniform on $[0, 1]$ and $u \cdot v$ the usual inner product in $\mathbb{R}^n$. Equation (6.4) now follows from the fact that all non trivial Fourier coefficients of a distribution uniform on the unit interval are zero.

Assume that $A = \{0\}^r \times [0, 1]^{n-r}$ is the lowest dimensional face of the unit hypercube that contains H. The first r coordinates of $\left(\sum U_i m_i\right)(\mathrm{mod}\, 1)$ and $\left(\sum U_i s_i\right)(\mathrm{mod}\, 1)$ then are identically zero while the vectors composed of the remaining $(n - r)$ coordinates have the same Fourier coefficients due to the previous paragraph and therefore the same distribution due to Proposition 2.1. As a constant random vector is independent of any other random vector it follows that the random vectors $\left(\sum U_i m_i\right)(\mathrm{mod}\, 1)$ and $\left(\sum U_i s_i\right)(\mathrm{mod}\, 1)$ have the same distribution. $\quad\square$

Corollary 1. *If H is generated by n linearly independent vectors (which does not necessarily mean that $H = \mathbb{Z}^n$; for example, H could be the group containing all integer valued vectors with even coordinates), the uniform distribution induced by H is a distribution uniform on $[0, 1]^n$.*

Proof. Follows from Proposition 6.6.b) and the fact that $H^\perp = \{0\}$. $\quad\square$

Corollary 2. *The s dimensional marginal distributions of U_H are uniform distributions induced by subgroups of $\mathbb{Z}^s$. In particular, all one dimensional marginal distributions are uniform on $[0, 1]$ if H is not equal to $\{0\}$.*

Proof. Follows from Proposition 6.6.a). $\quad\square$

Theorem 6.10 is one of the main results of this section. It gives necessary and sufficient conditions for weak-star convergence of $(A^k X)(\mathrm{mod}\,1)$ to the uniform distribution
induced by some subgroup of $\mathbb{Z}^n$. Its proof is based on the following two lemmas:

Lemma 6.7 *Let* $x_1, x_2, \ldots$ *denote a sequence of* n *dimensional vectors with complex
coordinates,* P *an* n *by* n *invertible matrix with complex coefficients and* $\|\cdot\|$ *any norm.
Then*

a) $\lim_{n\to\infty} \|x_n\| = +\infty$ *if and only if* $\lim_{n\to\infty} \|Px_n\| = +\infty$.

b) $\lim_{n\to\infty} \|x_n\| = 0$ *if and only if* $\lim_{n\to\infty} \|Px_n\| = 0$.

Proof. Follows from the fact that all norms are equivalent in finite dimensions, and that
$x \to \|Px\|$ defines a norm if P is invertible.   □

Lemma 6.8 *Let* P *be a* n *by* s *matrix with complex coefficients and denote its columns
by* $P_1, \ldots, P_s$. *Define the following sets of* n *dimensional complex valued vectors:*

$$\widehat{G} = \{m : P_i \cdot m = 0\,; \quad i = 1, \ldots, s\},$$

$$\widetilde{G} = \{m : P_i \cdot m = 0\,; \quad i = 2, \ldots, s\},$$

and given a complex number λ *consider the corresponding Jordan block matrix, that is,
a matrix with* λ *on its main diagonal, ones just below it and zeros elsewhere:*

$$J_\lambda \;=\; \begin{pmatrix} \lambda & 0 & \cdots & 0 & 0 \\ 1 & \lambda & \cdots & 0 & 0 \\ \vdots & \vdots & & \vdots & \vdots \\ 0 & 0 & \cdots & \lambda & 0 \\ 0 & 0 & \cdots & 1 & \lambda \end{pmatrix} \tag{6.5}$$

First assume $|\lambda| > 1$.

a) *If* $m \in \widehat{G}$ *then* $J'^k_\lambda P' m = 0$.

b) *If* $m \notin \widehat{G}$ *then* $\lim_{k\to\infty} \|J'^k_\lambda P' m\| = +\infty$.

Next suppose $|\lambda| = 1$.

c) *If* $m \in \widetilde{G}$ *then* $J'^k_\lambda P' m = (\lambda^k P_1 \cdot m, 0, \ldots, 0)$.

d) *If* $m \notin G$ *then* $\lim_{k\to\infty} \|J'^k_\lambda P' m\| = +\infty$.

Finally consider the case where $|\lambda| < 1$.

e) *For all* n *dimensional complex valued vectors* m: $\lim_{k\to\infty} \|J'^k_\lambda P' m\| = 0$.

Proof. Parts a) and c) are straight forward while e) follows from the fact that the norm
of J'^k_λ tends to zero as k tends to infinity.

To prove b) note that J'^k_λ is upper triangular with λ^k on its main diagonal and
$k\lambda^{k-1}$ just above it:

$$J'^k_\lambda = \begin{pmatrix} \lambda^k & k\lambda^{k-1} & 0 & \cdots & 0 & 0 \\ 0 & \lambda^k & k\lambda^{k-1} & \cdots & 0 & 0 \\ \vdots & \vdots & \vdots & & \vdots & \vdots \\ 0 & 0 & 0 & \cdots & \lambda^k & k\lambda^{k-1} \\ 0 & 0 & 0 & \cdots & 0 & \lambda^k \end{pmatrix}.$$

If $m \notin G$ then $P'm = (P_1 \cdot m, \ldots, P_r \cdot m, 0, \cdots, 0)'$ with $P_r \cdot m \neq 0$. The rth coordinate of $J'^k_\lambda P'm$ is then equal to $\lambda^k (P_r \cdot m)$ whose absolute value tends to infinity as k does.

To prove d) note that if $m \notin \widetilde{G}$ then $P'm = (P_1 m, \ldots, P_r m, 0, \cdots, 0)'$ with $P_r m \neq 0$ and $r \geq 2$. The $(r-1)^{\text{th}}$ component of $J'^k_\lambda P'm$ is equal to $\lambda^k (P_{r-1} \cdot m) + k\lambda^{k-1}(P_r \cdot m)$ whose absolute value tends to infinity as k does.   □

Lemma 6.9 *Assume the uniform distribution induced by the subgroup H of $\mathbb{Z}^n$ assigns no positive probability to the $(n-1)$ dimensional faces of the unit hypercube $[0,1]^n$, that is, to sets of the form $A_1 \times \cdots \times A_n$ with one A_i equal to $\{0\}$ or $\{1\}$ and the remaining A_j's equal to the unit interval. Let $b_1, b_2, \ldots$ be a sequence of vectors in $\mathbb{R}^n$ such that the distance between b_k and H tends to zero as k tends to infinity and $X_1, X_2, \ldots$ be a sequence of random vectors supported by $[0,1]^n$. Then $X_k(\mathrm{mod}\,1)$ converges to U_H in the weak-star topology as k tends to infinity if and only if $(X_k + b_k)(\mathrm{mod}\,1)$ does.*

Proof. Follows from Proposition 2.1, Lemma 4.1 and Proposition 6.6.b).   □

Corollary. *If the sequence of random vectors $X_1, X_2, \ldots$ supported by $[0,1]^n$ converges in the weak-star topology to a distribution uniform on $[0,1]^n$, U_n, then $(X_k + b_k)(\mathrm{mod}\,1)$ converges to U_n, where $b_1, b_2, \ldots$ denotes any (not necessarily convergent) sequence in $\mathbb{R}^n$.*

Proof. Follows from the fact that U_n assigns zero probability to the $(n-1)$ dimensional faces of $[0,1]^n$ and Lemma 6.9.   □

Theorem 6.10 *Let X be an n dimensional random vector with characteristic function tending to zero at infinity. Let A be an n by n matrix with Jordan form given by*

$$J = \begin{pmatrix} J_1 & 0 & \cdots & 0 \\ 0 & J_2 & \cdots & 0 \\ \vdots & \vdots & \ddots & \vdots \\ 0 & 0 & \cdots & J_r \end{pmatrix},$$

where the Jordan blocks $J_1, \ldots, J_r$ look like (6.5) and the corresponding eigenvalues $\lambda_1, \ldots, \lambda_k$ are not necessarily different.

Let P denote a matrix whose columns, $v_{1,1}, \ldots, v_{1,n_1}, \ldots, v_{r,1}, \ldots, v_{r,n_r}$, form a basis of generalized eigenvectors of A associated to J. That is, $A = PJP^{-1}$.

Define the following subgroups of $\mathbb{Z}^n$:

$\widetilde{G} = \{m \in \mathbb{Z}^n : m \text{ is orthogonal to all generalized eigenvectors associated to } \lambda_i\text{'s}$ with $|\lambda_i| \geq 1\}$.

$\widehat{G} = \{m \in \mathbb{Z}^n : m \text{ is orthogonal to all generalized eigenvectors associated to } \lambda_i\text{'s}$ with $|\lambda_i| > 1$ and all $v_{i,j}$'s associated to eigenvalues on the unit circle with $j \geq 2\}$.

Clearly $\widetilde{G} \subset \widehat{G}$.

a)   If $\widetilde{G} = \widehat{G}$ then $(A^k X)(\mathrm{mod}\, 1)$ converges weak-star to $U_{\widetilde{G}^{\perp}}$ as k tends to infinity.

The condition $\widetilde{G} = \widehat{G}$ is not only sufficient but also necessary for convergence in the following two senses:

b)   If $(A^k X)(\mathrm{mod}\, 1)$ converges and $\widetilde{G} \neq \widehat{G}$ then the limit has a distribution that depends on X.

c)   If the characteristic function of X, $\widehat{f}_X(\lambda)$, is different from zero for all $\lambda \in \mathbb{R}^n$, $\widetilde{G} = \widehat{G}$ is a necessary and sufficient condition for convergence of $(A^k X)(\mathrm{mod}\, 1)$ to the uniform distribution induced by some subgroup of $\mathbb{Z}^n$.

Proof. The proof is divided into three parts. In the first part, a) is proved for the case where $U_{\widetilde{G}^{\perp}}$ assigns zero probability to all $(n-1)$ dimensional faces of $[0,1]^n$. Part 2 completes the proof of a). Finally, b) and c) are proved in part 3.

1   First assume $m \in \widetilde{G}$.

Denote by P_i the matrix whose columns are $v_{i,1}, \ldots, v_{i,n_i}$. As

$$J'^k P' m = \begin{pmatrix} J'^k_1 P'_1 m \\ \vdots \\ J'^k_r P'_r m \end{pmatrix},$$

Lemmas 6.8 and 6.9 imply that $\|(P^{-1})' J'^k P' m\| = \|A'^k m\|$ tends to zero as k tends to infinity. Let $\widehat{f}_k(\lambda)$ denote the characteristic function of $(A^k X)(\mathrm{mod}\, 1)$. Lemma 4.1 implies that $\widehat{f}_k(2\pi m) = \widehat{f}_X(2\pi A'^k m)$ for every $m \in \mathbb{Z}^n$, and therefore $\widehat{f}_k(2\pi m)$ tends to one as k tends to infinity.

A similar argument shows that $\widehat{f}_k(2\pi m)$ tends to zero as k tends to infinity when $m \notin \widehat{G}$. Therefore, if $\widetilde{G} = \widehat{G}$, which is in particular the case when there are no eigenvalues on the unit circle, Propositions 2.1 and 6.5.b) and the fact that $\widetilde{G} = (\widetilde{G}^{\perp})^{\perp}$ imply that $(A^k X)(\mathrm{mod}\, 1)$ converges to $U_{\widetilde{G}^{\perp}}$.

2   Write

$$J = \begin{pmatrix} \widetilde{J}_1 & 0 \\ 0 & \widetilde{J}_2 \end{pmatrix},$$

with $\widetilde{J}_1$ and $\widetilde{J}_2$ corresponding to the eigenvalues outside or on the unit circle and inside the unit circle, respectively. Let s denote the number of columns of $\widetilde{J}_1$.

Assume, say, that $A = \{0\}^r \times [0,1]^{n-r}$ is the lowest dimensional face of the unit hypercube that contains $\widetilde{G}^{\perp}$. Then $e_1, \ldots, e_r$ belong to $\widetilde{G}$, where e_j denotes the j-th vector of the canonical basis of $\mathbb{R}^n$. This implies that the first s elements of the first r rows of P are equal to zero and therefore P may be written as

$$P = \begin{pmatrix} 0 & Q \\ R & S \end{pmatrix}.$$

Let Z_1 denote the first r coordinates of the random vector $P^{-1}X$ and Z_2 the random vector composed of the remaining $(n - r)$ coordinates. Then

$$A^k X = \begin{pmatrix} Q\tilde{J}_2^k Z_2 \\ R\tilde{J}_1^k Z_1 + S\tilde{J}_2^k Z_2 \end{pmatrix}.$$

It follows that the first r coordinate of $(A^k X)(\mathrm{mod}\,1)$ converge to a distribution with all its mass at the origin. Conditioning on the value of Z_2 and applying Proposition 2.3 and Lemma 6.9 reduces the problem of determining the limit of the remaining $(n - r)$ coordinates to that of studying the limit of $(R\tilde{J}_1^k Z_1)(\mathrm{mod}\,1)$ and this can be done using the same argument as in part 1. Finally, it is easy to see that the limit obtained in this way for $(A^k X)(\mathrm{mod}\,1)$ corresponds to the uniform distribution induced by $\tilde{G}^{\perp}$.

3   Assume there exists $m \in \mathbf{Z}^n$ which belongs to $\hat{G}$ and does not belong to $\tilde{G}$. For the remainder of the proof let $\|x\|$ denote the sum of the absolute values of the coordinates of $P'x$. Lemma 6.7 implies that $\|A'^k m\|$ converges to a finite, strictly positive constant c. For $m \in \mathbf{Z}^n$, $\hat{f}_k(2\pi m) = \hat{f}_X(2\pi A'^k m)$, and hence if $(A^k X)(\mathrm{mod}\,1)$ converges, the limiting distribution's Fourier coefficient associated to m depends on the values the characteristic function of X takes on the circle (in the special norm being considered) of radius c. This proves part b).

To prove part c) note that, from the previous paragraph it follows that, for sufficiently large k, $A'^k m$ belongs to the set $\mathcal{C}$ of points whose distance to the origin is between $c/2$ and $2c$. This set is compact and characteristic functions are continuous, and therefore the function $\lambda \to |\hat{f}(\lambda)|$, restricted to $\mathcal{C}$, attains its maximum and minimum values on $\mathcal{C}$. This value cannot be zero by hypothesis and cannot be one because, as $\mathcal{C}$ does not contain the origin, it would imply that there exists a non trivial linear combination of the X_i's having a lattice distribution, contradicting the hypothesis that $\hat{f}(\lambda)$ vanishes at infinity. It follows that the Fourier coefficient associated to m cannot converge to zero or one and therefore $(A^k X)(\mathrm{mod}\,1)$ does not converge to the uniform distribution induced by any subgroup of $\mathbf{Z}^n$.   □

Remark 1. Let $G = \{m \in \mathbf{Z}^n : m$ is orthogonal to all generalized eigenvectors associated to λ_i's with $|\lambda_i| > 1\}$.

If A does not have eigenvalues on the unit circle then $\tilde{G} = \hat{G} = G$. In this case $(A^k X)(\mathrm{mod}\,1)$ converges to the uniform distribution induced by G. The limit is uniform on $[0,1]^n$ if and only if $G = \{0\}$, that is, if and only if there does not exist a non trivial integer valued vector orthogonal to all generalized eigenvectors associated to eigenvalues outside the unit circle.

Remark 2. In general, $(A^k X)(\mathrm{mod}\,1)$ converges to a distribution uniform on the unit hypercube if $\hat{G} = \{0\}$.

Remark 3. If all eigenvalues have absolute value larger than one then $G = \{0\}$ and $(A^k X)(\mathrm{mod}\,1)$ converges to a distribution uniform on $[0,1]^n$, as may be inferred from Theorem 6.2. On the other hand, if all eigenvalues lie inside the unit circle, A^k tends to the null matrix and therefore $(A^k X)(\mathrm{mod}\,1)$ converges almost surely to a distribution with all its mass at the origin.

Remark 4. From standard linear algebra (see e.g. Hofmann and Kunze, 1971, Chap. 7) it follows that

$$v_{i,j} = (A - \lambda_i I)^{j-1} v_{1,j}, \tag{6.6}$$

where I denotes the n by n identity matrix.

Assume $J_1, \ldots, J_p$ are all the Jordan blocks associated to the eigenvalue λ and let $s = \sum_{k=1}^p n_k$. Then $\{v_{1,1}, \ldots, v_{1,n_1}, \ldots, v_{p,1}, \ldots, v_{p,n_p}\}$ form a basis for the kernel of $(A - \lambda I)^s$. This fact combined with (6.6) means that the $(A - \lambda I)$-cyclic subspace generated by $v_{1,1}, v_{2,1}, \ldots, v_{j,1}$ is equal to the kernel of $(A - \lambda I)^s$. It is in this sense that the $v_{i,1}$'s have a special role compared to that of the $v_{i,j}$'s with $j > 1$.

Remark 5. The matrix $A = \begin{pmatrix} 1 + 2\sqrt{2} & -4 \\ 2 & 1 - 2\sqrt{2} \end{pmatrix}$ has both eigenvalues equal to one. The corresponding Jordan matrix, J, and matrix with columns equal to a basis of generalized eigenvectors, P, are

$$J = \begin{pmatrix} 1 & 0 \\ 1 & 1 \end{pmatrix} \qquad ; \qquad P = \begin{pmatrix} 1 & \sqrt{2} \\ \sqrt{2}/4 & 1 \end{pmatrix}.$$

Therefore $\widetilde{G} = \widehat{G} = \{0\}$ and $(A^k X)(\mathrm{mod}\,1)$ converges to a distribution uniform on the unit square despite the fact that none of its eigenvalues lies outside the unit circle.

Remark 6. A simple case where the limit of $(A^k X)(\mathrm{mod}\,1)$ depends on the distribution of X is when A is equal to the n by n identity matrix, or, more generally, if

$$A = \begin{pmatrix} B & 0 \\ 0 & 1 \end{pmatrix},$$

where B denotes an $(n-1)$ by $(n-1)$ matrix with all its eigenvalues outside the unit circle and both zeros denote vectors of zeros.

Remark 7. If the matrix A has at least one eigenvalue outside the unit circle and no eigenvalue on the unit circle, $\widehat{G}$ has dimension larger or equal than one and therefore the limit of $(A^k X)(\mathrm{mod}\,1)$ is not a constant random vector.

Remark 8. That a dynamical system's behavior should depend on whether certain quantities are rational or irrational is a surprising fact that often appears in mathematical models for dynamical systems. It is known as the problem of "small divisors". For an introduction to this topic, see Moser (1973).   □

Theorem 6.10 shows that for every subgroup H of $\mathbb{Z}^n$ there exists a non empty family of matrices $\mathcal{A}_H$ such that for every matrix A in $\mathcal{A}_H$ the random vector $(A^k X)(\mathrm{mod}\,1)$

converges to the uniform distribution induced by H as k tends to infinity. In Proposition 6.11 it is proved that if A is chosen randomly from a density on the set of n by n matrices, among all uniform distributions induced by subgroups of $\mathbb{Z}^n$ only two have a positive probability of occurring: a distribution uniform on $[0,1]^n$ and a distribution with all its mass at the origin.

Proposition 6.11 *Assume A is a an n by n random matrix and X an n dimensional random vector , and suppose (X, A) has an absolutely continuous distribution on $\mathbb{R}^n \times \mathbb{R}^{n^2}$. Define $\mathcal{V}$ as the set of all matrices with all their eigenvalues inside the unit circle and let $p = \Pr\{A \in \mathcal{V}\}$. Denote a distribution uniform on $[0,1]^n$ by U_n and a distribution with all its mass at the origin by 0. Then, as k tends to infinity, $(A^k X)(\mathrm{mod}\, 1)$ converges to a distribution equal to 0 with probability p and equal to U_n with probability $(1-p)$.*

Proof. From the fact that A has a density it follows that if $A \notin \mathcal{V}$, the subgroup $\widehat{G}$ defined in Theorem 6.10 is equal to $\{0\}$ with probability one: all eigenvalues have multiplicity one and do not lie on the unit circle, except for a set of Lebesgue measure zero. Further, the coordinates of every eigenvector are linearly independent over the rationals with probability one. Therefore, denoting by $\mathcal{U}$ the set of matrices A for which $(A^k X)(\mathrm{mod}\, 1)$ conditioned on the value of A converges to U_n, it has just been showed that $\Pr\{A \in \mathcal{U}\} = 1 - p$.

Apply Proposition 2.3 to conclude that $(A^k X | A \in \mathcal{V})(\mathrm{mod}\, 1)$ converges to 0 while $(A^k X | A \in \mathcal{U})(\mathrm{mod}\, 1)$ converges to U_n. These two facts combined with the previous paragraph complete the proof. $\square$

Theorem 6.10 shows that – with few exceptions – the random vector $(A^k X)(\mathrm{mod}\, 1)$ converges to a distribution that does not depend on X. This provides a mathematical formulation of the intuitive idea of unpredictability of a system whose position at time k is $(A^k X)(\mathrm{mod}\, 1)$, where the random vector X describes initial conditions. It is equivalent to having the one dimensional distributions of the infinite dimensional random sequence

$$Y_k = \left((A^k X)(\mathrm{mod}\, 1), (A^{k+1} X)(\mathrm{mod}\, 1), (A^{k+2} X)(\mathrm{mod}\, 1), \ \ldots \ \right)$$

converge to a limit that does not depend on the distribution of X. A much stronger notion of unpredictability applies if the stochastic process defined by Y_k has a limit that does not depend on the distribution of X. The following two propositions show that, quite generally, this is the case.

Proposition 6.12 *Assume X is an n dimensional random vector and A an n by n matrix satisfying the assumptions and notation introduced in Theorem 6.10. For a fixed positive integer r let*

$$Z_k = \left((A^k X)(\mathrm{mod}\, 1), (A^{k+1} X)(\mathrm{mod}\, 1), \ldots, (A^{k+r-1} X)(\mathrm{mod}\, 1) \right).$$

Given $m \in \mathbb{Z}^{nr}$ write $m = (m_1, \ldots, m_r)'$ with $m_i \in \mathbb{Z}^n$, and define the following subgroups of $\mathbb{Z}^{nr}$:

$$\widehat{G} = \{ m \in \mathbb{Z}^{nr} \colon \textstyle\sum_{i=0}^{r-1} J'^i P' m_i \text{ has all coordinates associated to generalized eigen-}$$
vectors with $|\lambda_i| \geq 1$ equal to zero $\}$.

$\widehat{G} = \{m \in \mathbb{Z}^{nr}: \sum_{i=0}^{r-1} J'^i P' m_i$ has all coordinates associated $v_{i,j}$'s with $|\lambda_i| > 1$ and all coordinates corresponding to $v_{i,j}$'s with $|\lambda_i| = 1$ and $j \geq 2$ equal to zero$\}$.

If $\widetilde{G} = \widehat{G}$ then, as k tends to infinity, Z_k converges in the weak-star topology to $U_{\widetilde{G}^\perp}$. In particular, if $\widehat{G} = \{0\}$, Z_k converges to a distribution uniform on $[0,1]^n$.

Proof. Due to Lemma 4.1, the Fourier coefficient of Z_k associated to the vector $m = (m_1, \ldots, m_r)' \in \mathbb{Z}^{nr}$ is equal to the characteristic function of X evaluated at the vector $2\pi(P^{-1})' J'^k \sum_{i=0}^{r-1} J'^i P' m_i$, so the proof of Theorem 6.10 carries through by letting $\sum_{i=0}^{r-1} J'^i P' m_i$ play the role of $P'm$.

Remark 1. Given $m = (m_1, \ldots, m_r)' \in \mathbb{Z}^{nr}$ let $m(\lambda) = \sum_{i=0}^{r-1} \lambda^i m_i$. That is, the function $m(\lambda)$ associates to a complex number λ an n dimensional vector whose components are polynomials of degree less than or equal to $(r-1)$ with integer coefficients.

The subgroups $\widetilde{G}$ and $\widehat{G}$ are usually difficult to interpret, yet if A is diagonalizable then $\widehat{G} = \{m \in \mathbb{Z}^{nr} : m(\lambda_i) \cdot v_{i,j} = 0$ for all $v_{i,j}$'s associated to eigenvalues outside the unit circle and all $v_{i,j}$'s with corresponding eigenvalues on the unit circle and $j \geq 2\}$.

Remark 2. Remark 1 implies that a sufficient condition for weak-star convergence of Z_k to a distribution uniform on $[0,1]^{nr}$ is that the coordinates of every generalized eigenvector $v_{i,j}$ associated to an eigenvalue which is not inside the unit circle be linearly independent over the field generated by rational linear combinations of $1, \lambda_i, \ldots, \lambda_i^{r-1}$.

Remark 3. Assume the kth row of A, a_k, has integer (or rational) coefficients and that A is diagonalizable with eigenvalues $\lambda_1, \ldots, \lambda_n$ and corresponding basis of eigenvectors $v_1, \ldots, v_n$. The fact that $(A - \lambda_i I)v_i = 0$, $i = 1, \ldots, n$; implies that $(a_k - \lambda_i e_k) \cdot v_i = 0$, $i = 1, \ldots, n$; where e_k denotes the kth vector of the canonical basis of $\mathbb{R}^n$. Remark 2 then implies that Z_k does not converge to a distribution uniform on $[0,1]^{nr}$ if $r \geq 2$. In particular, if A has no eigenvalues on the unit circle, $(A^k X, A^{k+1} X)(\text{mod } 1)$ always has a limit that is the uniform distribution induced by a subgroup of $\mathbb{Z}^{2n}$ of dimension smaller than $2n$.

This remark may be extended as follows to the case where A is a matrix with integer (or rational) coefficients that is not diagonalizable. With the notation introduced in Theorem 6.10, let $m = \max n_i$ and $m_k(\lambda)$ be the k-th row of $(A - \lambda I)^m$. Remark 4 following Theorem 6.10 and the fact that the v_{i,n_i}'s are eigenvectors of A imply that $m_k(\lambda_i) \cdot v_{i,j} = 0$ for all i and j and therefore the random vector $((A^k X)(\text{mod } 1), \ldots, (A^{k+r-1} X)(\text{mod } 1))$ cannot converge to a distribution uniform on $[0,1]^{nr}$ if r is larger than m.

Proposition 6.13 *Let A denote an n by n matrix with no eigenvalue on the unit circle and X an n dimensional random vector with characteristic function vanishing at infinity. Then the infinite dimensional stochastic process*

$$Y_k = ((A^k X)(\text{mod } 1), (A^{k+1} X)(\text{mod } 1), (A^{k+2} X)(\text{mod } 1), \ldots)$$

converges weak-star (with respect to the topology generated by the cylinder sets) to a stochastic process Y on $[0,1]^\infty$ whose distribution does not depend on that of X and whose s dimensional marginals are uniform distributions induced by subgroups of $\mathbb{Z}^s$. Further, if A has at least one eigenvalue outside the unit circle, all one dimensional marginal distributions of Y are uniform on $[0,1]$.

Proof. As the Y_k's take values in $[0,1]^\infty$ which is compact (Tychonoff's Theorem), they form a tight sequence. Therefore (Loève, 1977, p.195) the Y_k's are relatively compact, that is, every subsequence has a subsequence that converges. Due to Proposition 6.12 the limiting process has finite dimensional distributions that do not depend on the particular sequence. The topology being considered is such that a process is determined by its finite dimensional distributions and hence Y_k converges to a process Y, whose s dimensional marginals are uniform distributions induced by subgroups of $\mathbb{Z}^s$ (see Proposition 6.12 and Corollary 2 of Proposition 6.6). The limiting process has one dimensional marginals uniform on $[0,1]$ due to Corollary 2 of Proposition 6.6.   □

Remark. An argument similar to that used in the proof of Proposition 6.11 combined with Remark 2 following Proposition 6.12 implies that if X and A have a joint density on $\mathbb{R}^n \times \mathbb{R}^{n^2}$ then Y_k converges to a distribution uniform on $[0,1]^\infty$ with probability equal to the probability that at least one eigenvalue of A be outside the unit circle, and zero otherwise.   □

The study of $(e^{tA}X)(\mathrm{mod}\,1)$ as the real number t tends to infinity follows the same lines as what was just done for $(A^k X)(\mathrm{mod}\,1)$. A result similar to Lemma 6.8 leads to the following theorem, whose proof is omitted.

Theorem 6.14 *Assume X is an n dimensional random vector and A an n by n matrix satisfying the hypothesis and notation introduced in Theorem 6.10. Denote the real part of the complex number λ by $\mathrm{Re}(\lambda)$ and define the following subgroups of $\mathbb{Z}^n$:*

$\tilde{H} = \{m \in \mathbb{Z}^n : m$ *is orthogonal to all generalized eigenvectors associated to λ_i's with* $\mathrm{Re}(\lambda_i) \geq 0\}$.

$\hat{H} = \{m \in \mathbb{Z}^n : m$ *is orthogonal to all generalized eigenvectors associated to λ_i's with* $\mathrm{Re}(\lambda_i) > 0$ *and all $v_{i,j}$'s associated to eigenvectors with $\mathrm{Re}(\lambda_i) = 0$ and $j \geq 2\}$.*

a)   *If $\tilde{H} = \hat{H}$, which in particular is the case if A has no purely imaginary eigenvalues, $(e^{tA}X)(\mathrm{mod}\,1)$ converges in the weak-star topology to $U_{\tilde{H}^\perp}$ as t tends to infinity. In particular, if $\tilde{H} = \{0\}$, $(e^{tA}X)(\mathrm{mod}\,1)$ converges to U_n.*

The condition $\tilde{H} = \hat{H}$ is not only sufficient but also necessary for convergence in the following two senses:

b)   *If $\tilde{H} \neq \hat{H}$ and $(e^{tA}X)(\mathrm{mod}\,1)$ converges, the limit has a distribution that depends on X.*

c)   *If the characteristic function of X, $\hat{f}_X(\lambda)$, evaluated at any λ in $\mathbb{R}^n$ differs from zero, $\hat{H} = \tilde{H}$ is necessary and sufficient for convergence of $(e^{tA}X)(\mathrm{mod}\,1)$ to a uniform* *distribution induced by some subgroup of* $\mathbb{Z}^n$. *In particular,* $(e^{tA}X)(\mathrm{mod}\,1)$ *converges to a distribution uniform on the unit hypercube if and only if* $\widehat{H} = \{0\}$.

6.1.3 Rates of Convergence

Having studied conditions under which $(A^k X)(\mathrm{mod}\,1)$ and $(e^{tA}X)(\mathrm{mod}\,1)$ converge to a uniform distribution in the weak-star topology, the problem of finding bounds on the rate of convergence is now addressed.

The two dimensional case is considered. The Kolmogorov distance between two random vectors X, Y with distributions F and G is defined as

$$d_K(X,Y) = \sup_{x \in \mathbb{R}^2} |F(x) - G(x)|.$$

One caveat of the Kolmogorov distance is that it is not translation invariant, i.e. the Kolmogorov distance between $X(\mathrm{mod}\,1)$ and $Y(\mathrm{mod}\,1)$ is usually different from that between $(X + b)(\mathrm{mod}\,1)$ and $(Y + b)(\mathrm{mod}\,1)$, where b is a fixed vector in $\mathbb{R}^n$. The bounds derived in this section do not have this problem: they are translation invariant in the sense described above.

From Propositions 2.2 it follows that $(A^k X)(\mathrm{mod}\,1)$ and $(e^{tA}X)(\mathrm{mod}\,1)$ converge to a distribution uniform on the unit square, U_2, if and only if $d_K((A^k X)(\mathrm{mod}\,1),\, U_2)$ and $d_K((e^{tA}X)(\mathrm{mod}\,1),\, U_2)$ tend to zero, respectively.

Upper bounds on the Kolmogorov distance between $(AX)(\mathrm{mod}\,1)$ and U_2 are derived in Proposition 6.16. They are used to bound the rate at which $(A^k X)(\mathrm{mod}\,1)$ tends to U_2. These bounds are useful when one eigenvalue of A has absolute value larger than one while the other eigenvalue has absolute value less than or equal to one. They depend on number theoretic properties of the coordinates of the eigenvector of A corresponding to the eigenvalue outside the unit circle. Bounds for the case where both eigenvalues have absolute value larger than one were obtained in Theorem 6.2.

A similar approach is used to find upper bounds for the Kolmogorov distance between $(e^{tA}X)(\mathrm{mod}\,1)$ and U_2.

The problem of finding an upper bound for $d_K((AX)(\mathrm{mod}\,1),\, U_2)$ can be reduced to that of finding an upper bound for the Kolmogorov distance between $(X, uX)'(\mathrm{mod}\,1)$ and U_2, where u is an irrational number, and this is done in Proposition 6.15. The result uses the notion of type of an irrational number (see Chap. 2.5). Recall that the set of irrational numbers with type larger than one has Lebesgue measure zero.

Proposition 6.15 *Assume X is a random variable with bounded total variation* $\mathrm{V}(X)$ *and u an irrational number of type less than 2. Then there exists a constant that only depends on u, $c(u)$, such that*

$$d_K((X, uX)'(\mathrm{mod}\,1),\, U_2) \le c(u)\mathrm{V}(X).$$

Proof. Denote the characteristic function of X by $\widehat{f}_X(\lambda)$ and that of $Y = (X, uX)'$ by $\widehat{f}_Y(\lambda_1, \lambda_2)$. Berry-Esseen type bounds for the Kolmogorov distance between two n dimensional distributions with support contained in $[0,1]^n$ are derived by Niederreiter and Philipp (1973). These bounds lead to

$$d_K\left((X, uX)'(\mathrm{mod}\,1), U_2\right) \le \frac{32.5}{\pi} \sum_{(k,l)\ne(0,0)} \frac{|\widehat{f}_Y(2\pi k, 2\pi l)|}{|kl|}.$$

Integration by parts shows that $|\widehat{f}_X(\lambda)| \le V(X)/\lambda$ and a calculation from first principles yields $\widehat{f}_Y(\lambda_1, \lambda_2) = \widehat{f}_X(2\pi(k + lu))$. Therefore

$$d_K\left((X, uX)'(\mathrm{mod}\,1),\, U_2\right)$$

$$\le \frac{32.5 V(X)}{2\pi^2} \left\{ \sum_{k\ne0} \frac{1}{k^2} + \frac{1}{|u|} \sum_{l\ne0} \frac{1}{l^2} + \sum_{k\ne0, l\ne0} \frac{1}{|kl||k + lu|} \right\}.$$

The proof now follows from Proposition 2.13. □

Proposition 6.16 *Assume A is a two by two matrix with eigenvalues λ and μ and corresponding eigenvectors $(1, u)$ and $(s, 1)$. Suppose u is irrational of type less than 2. Let (X_1, X_2) be a two dimensional random vector such that the total variation of $X_1 - sX_2$ conditioned on $X_2 - uX_1$ is finite and denote it by V_0. Then*

$$d_K\left((AX)(\mathrm{mod}\,1),\, U_2\right) \le c(u)|1 - us|V_0 / |\lambda|,$$

where $c(u)$ is the constant defined in Proposition 6.15. Therefore:

$$d_K\left((A^k X)(\mathrm{mod}\,1),\, U_2\right) \le c(u)|1 - us|V_0 |\lambda|^{-k}$$

and

$$d_K\left((e^{tA} X)(\mathrm{mod}\,1),\, U_2\right) \le c(u)|1 - us|V_0 e^{-|\lambda|t}.$$

Proof. Letting $Z_1 = (X_1 - sX_2)/(1 - us)$ and $Z_2 = (X_2 - uX_1)/(1 - us)$, and using the fact that $A = PDP^{-1}$, where the columns of P are the eigenvectors of A and D is a diagonal matrix with the corresponding eigenvalues on its diagonal, leads to

$$AX = \lambda \begin{pmatrix} Z_1 \\ uZ_1 \end{pmatrix} + \mu \begin{pmatrix} sZ_2 \\ Z_2 \end{pmatrix}.$$

Conditioning on the value of Z_2, applying Proposition 6.15 and basic properties of the total variation of a random variable (see Proposition 3.8) and Proposition 2.4 to uncondition, leads to

$$d_K\left((AX)(\mathrm{mod}\,1),\, U_2\right) \le c(u)|1 - us|V_0 / |\lambda|.\,□$$

Remark 1. Lemma 6.1 implies that if X_1 and X_2 have bounded variation when conditioned on each other then $V_0 = V_1(X_1 - sX_2, X_2 - uX_1)$ is finite and the following inequalities hold:

$$V_0 \leq \frac{1+|u|}{|1-us|} \max\left(V_1(X_1,X_2), V_1(X_2,X_1)\right),$$

$$V_0 \leq \frac{1+|s|}{|1-us|} \{V_1(X_1,X_2) + V_!(X_2,X_1)\}.$$

Remark 2. The approach leading to Proposition 6.16 could be extended to higher dimensions if a result analogous to Proposition 2.13 were proved. Consider, for example, the three dimensional case. If $(1,u_1,u_2)$ is the eigenvector associated to the eigenvalue λ with $|\lambda| > 1$ and

$$\sum_{(k,l,m)\neq(0,0,0)} \frac{1}{|klm||k+lu_1+mu_2|} < \infty,$$

the higher dimensional version of the result due to Niederreiter and Philipp (1973) implies that

$$d_K\left((X,u_1X,u_2X)', U_3\right) \leq c(u_1,u_2)V(X),$$

and the argument given in the proof of Proposition 6.16 can be used to derive upper bounds for $d_K\left((A^kX)(\mathrm{mod}\,1), U_2\right)$ and $d_K\left((e^{tA}X)(\mathrm{mod}\,1), U_2\right)$.

Remark 3. Let b denote a vector in $\mathbb{R}^n$. The fact that the mean-conditional variation is invariant under translations implies that all bounds derived above apply if $X+b$ is considered instead of X. For example, combining Proposition 6.16 and Remark 1:

$$\sup_{b\in\mathbb{R}^n} d_K\left((AX)(\mathrm{mod}\,1), U_2\right)$$
$$\leq c(u)(1+|u|)\max\left(V_1(X_1,X_2), V_1(X_2,X_1)\right).\square \qquad (6.7)$$

Example. Propositions 2.13 and 2.14 show that the constant $c(u)$ can be computed explicitly if u is a quadratic irrational. Consider, for example, $A^kX(\mathrm{mod}\,1)$ when

$$A = \begin{pmatrix} 1 & 1 \\ 1 & 2 \end{pmatrix}.$$

Then

$$A^k = \begin{pmatrix} f_{2k-2} & f_{2k-1} \\ f_{2k-1} & f_{2k} \end{pmatrix},$$

where the f_k's denote the Fibonacci numbers: $f_0 = f_1 = 1$ and $f_n = f_{n-1} + f_{n-2}$ for $n \geq 2$.

The eigenvalues of A are $(3+\sqrt{5})/2$ and $(3-\sqrt{5})/2$ with corresponding eigenvectors $(1,u)$ and $(s,1)$, where $u = (1+\sqrt{5})/2$ and $s = -u$. From the example at the end of Sect. 2.5 it follows that $c(u) \leq 276$. This and (6.7) imply that

$$d_K\left(A^kX(\mathrm{mod}\,1), U_2\right) \leq 723\,M(X)\left(\frac{3+\sqrt{5}}{2}\right)^{-k}, \qquad (6.16)$$

where $M(X) = \max\left(V_1(X_1,X_2), V_1(X_2,X_1)\right)$. Assume X is such that $M(X)$ is less than three times the corresponding expression for a distribution uniform on $[0,1]^2$. Then

(6.16) implies that the Kolmogorov distance between $A^k X(\mathrm{mod}\,1)$ and a distribution uniform on the unit square is less than 0.29 after 10 iterations, less than 0.0024 after 15 iterations and less than 2×10^{-5} after 20 iterations.

6.2 Linear Differential Equations

Consider an nth order, homogeneous, linear differential equation with constant coefficients:

$$x^{(n)}(t) + a_1 x^{(n-1)}(t) + \ldots + a_{n-1} x'(t) + a_n x(t) = 0 \qquad (6.9)$$

and vector of initial conditions equal to

$$y_0 = \left(x(0), x'(0), \ldots, x^{(n-1)}(0)\right)'. \qquad (6.10)$$

Linear differential equations are used to model various phenomena in physics, biology and economics. From a mathematical point of view they are very simple models and even though the processes under consideration are often non linear, a linear approximation may be a good starting point before a more sophisticated analysis is undertaken.

Let $x(t)$ denote the unique solution to (6.9) satisfying initial conditions given by (6.10). The behavior of $x(t)(\mathrm{mod}\,1)$ as t tends to infinity is studied in this section. There are two possible sources of randomness in the behavior of $x(t)$: the coefficients $a_1, \ldots, a_n$ in the differential equation (6.9) and the initial conditions specified in (6.10). The a_i's are assumed fixed while y_0 is supposed random. If λ is a (real or complex) root of the characteristic equation associated to (6.9):

$$z^n + a_1 z^{n-1} + \ldots + a_{n-1} z + a_n = 0, \qquad (6.11)$$

the solution, $x(t)$, has a term of the form $c e^{\lambda t}$. It therefore seems reasonable to expect that $x(t)(\mathrm{mod}\,1)$ converges to a distribution uniform on the unit interval if (6.11) has at least one root with positive real part. This statement is made precise in the following proposition.

Proposition 6.17 *Let $x(t)$ denote the unique solution of (6.9) satisfying initial conditions given in (6.10). Assume the a_i's are fixed and y_0 is described by an n dimensional density such that every coordinate has bounded variation when conditioned on the random vector composed of the remaining $(n-1)$ coordinates.*

Among the roots of (6.11) with largest real part, denote by λ that with largest multiplicity, q.

Define $M(y_0) = \max_{1 \le i \le n} V_i(y_0)$.

Then

$$d_{\mathrm{V}}\left(x(t)(\mathrm{mod}\,1),\, U\right) \le \frac{k}{8} M(y_0) t^{-q+1} e^{-(\mathrm{Re}\lambda)t}, \qquad (6.12)$$

where

$$k = \begin{cases} \sum_{j=0}^{n-q} \frac{(j+q-1)!}{j!} |\lambda|^j, & \text{if } \lambda \text{ is real,} \\[2mm] \sqrt{2} \sum_{j=0}^{n-q} \frac{(j+q-1)!}{j!} |\lambda|^j, & \text{otherwise.} \end{cases} \tag{6.13}$$

Proof. Let $\lambda_1, \ldots, \lambda_m$ denote the distinct real roots of (6.11) with corresponding multiplicities $n_1, \ldots, n_m$ and $\mu_1, \bar{\mu}_1, \ldots, \mu_r.\bar{\mu}_r$ the distinct complex roots with multiplicities $\eta_1, \ldots, \eta_r$. Then

$$x(t) = \sum_{i=1}^{m} y_i(t) + \sum_{i=1}^{r} z_i(t),$$

where

$$y_i(t) = \Big(\sum_{j=0}^{n_i-1} c_{i,j} t^j \Big) e^{\lambda_i t},$$

$$z_i(t) = \sum_{j=0}^{\eta_i-1} t^j \big(v_{i,j} \cos(\omega_i t) + w_{i,j} \sin(\omega_i t) \big) e^{r_i t},$$

with $\mu_j = r_j e^{i\omega_j}$.

Let

$$c = (c_{1,0}, \ldots, c_{1,n_1-1}, \ldots, c_{m,0}, \ldots, c_{m,n_m-1}, v_{1,0}, w_{1,0}, \ldots, v_{r,\eta_r-1}, w_{r,\eta_r-1})',$$

and

$$s(t) = \big(e^{\lambda_1 t}, \ldots, t^{n_1-1} e^{\lambda_1 t}, \ldots, e^{\lambda_m t}, \ldots, t^{n_m-1} e^{\lambda_m t}, \cos(\omega t), \sin(\omega t), \ldots,$$

$$t^{\eta_r-1} \cos(\omega_r t), t^{\eta_r-1} \sin(\omega_r t) \big).$$

Then $x(t) = c \cdot s(t)$ and it follows that $y_0 = Bc$, where the kth row of B is equal to $s^{(k-1)}(0)$. Hence $x(t) = s(t)' B^{-1} y_0$, with $s(t)'$ denoting the corresponding transpose. Proposition 6.4 and Lemma 6.1 imply that

$$\mathrm{dv}\big(x(t)(\mathrm{mod}\,1),\, U\big) \leq \frac{M(y_0)}{8} \min_{1 \leq j \leq n} \frac{\sum_{i=1}^{n} |b_{i,j}|}{|s_j(t)|}, \tag{6.14}$$

where $b_{i,j}$ denotes the (i,j)-th element of B and $s_j(t)$ the jth coordinate of $s(t)$.

If λ is real, consider j such that $s_j(t) = t^{q-1} e^{\lambda t}$ to obtain an upper bound on the right hand side of (6.14). Straightforward calculus shows that $s_j^{(k)}(0)$ is equal to $k! \lambda^{k-q+1}/(k-q+1)!$ if $k \geq q-1$ and zero otherwise. Therefore $\sum_i |b_{i,j}| \leq k$ with k defined in (6.13) and the required conclusion follows.

If $\lambda = re^{i\omega}$ is complex, consider j and $j+1$ such that $s_j(t) = t^{q-1} e^{rt} \cos(\omega t)$ and $s_{j+1}(t) = t^{q-1} e^{rt} \sin(\omega t)$, to obtain an upper bound for the right hand side of (6.14). An argument similar to the one given in the previous paragraph shows that both $\sum_i |b_{i,j}|$ and $\sum_i |b_{i,j+1}|$ are bounded by $k/\sqrt{2}$. This combined with the fact that the minimum between $1/|\cos x|$ and $1/|\sin x|$ is $\sqrt{2}$ completes the proof. $\qquad\square$

Remark 1. Proposition 6.17 implies that if at least one root of (6.11) has positive real part or if a purely imaginary root has multiplicity larger than one, then $x(t)(\mathrm{mod}\,1)$ converges in the variation distance to a distribution uniform on the unit interval as t tends to infinity.

Remark 2. If all solutions of (6.11) are inside the unit circle, $x(t)$ tends to zero as t tends to infinity.

Remark 3. Proposition 6.17 can be used to obtain upper bounds for the variation distance between $x(t)(\mathrm{mod}\,1)$ and U when both initial conditions and coefficients of the differential equation are random. Assume the a_i's in (6.9) and y_0 have a joint density. With probability one the multiplicity of every root of (6.11) is one and therefore the bound in (6.12) only depends on the maximum among the real parts of the roots of (6.11), say $\mathrm{Re}(\lambda)$. Let $p = \mathrm{Pr}\{\mathrm{Re}(\lambda) > 0\}$. Then $x(t)(\mathrm{mod}\,1)$ converges in the weak-star topology to a distribution that is uniform on the unit interval with probability p and identically zero otherwise. Further, if $p = 1$, Propositions 6.17 and 2.8.a) provide useful bounds for the variation distance between $x(t)(\mathrm{mod}\,1)$ and U:

$$d_V\left(x(t)(\mathrm{mod}\,1),\,U\right) \;\leq\; \frac{\sqrt{2}}{8}\mathrm{E}\left\{\max_i V_i(y_0|\mathrm{Re}\lambda)e^{-(\mathrm{Re}\lambda)t}\right\},$$

where the expected value is taken over the distribution of $\mathrm{Re}(\lambda)$. In most cases the coefficients of the differential equation and the initial conditions described by y_0 may be assumed independent. Then

$$d_V\left(x(t)(\mathrm{mod}\,1),\,U\right) \;\leq\; \frac{\sqrt{2}}{8}M(y_0)\mathrm{E}\left(e^{-(\mathrm{Re}\lambda)t}\right),$$

with $M(y_0) = \max_i V_i(y_0)$. The rate of convergence therefore depends on the behavior of the moment generating function of $\mathrm{Re}(\lambda)$ at minus infinity.

Remark 4. From Remark 2 it follows that the only case of interest not covered by Proposition 6.17 is when (6.11) has some roots with multiplicity one on the unit circle $\mu_1, \bar{\mu}_1, \ldots, \mu_p.\bar{\mu}_p$, and the remaining roots have absolute value less than one. Then

$$x(t) = \sum_{i=1}^{p} r_i \cos(\omega_i t + \phi_i) + x_0(t),$$

where the r_i's and ϕ_j's depend on the coefficients of the differential equation and the vector of initial conditions, y_0, and $x_0(t)$ tends to zero as t tends to infinity.

If uncertainty about unknown quantities is described by a joint density on the random vector $(\omega_1, \ldots, \omega_p, y_0)$, Theorem 4.4 implies that $x(t)$ converges in the weak-star topology to a linear combination of independent, identically distributed arcsine laws, $\sum c_i S_i$, with the S_i's having common density

$$f(s) \;=\; \frac{1}{\pi\sqrt{1-s^2}} \qquad\qquad ;-1 < s < 1, \qquad\qquad (6.15)$$

and the c_i's independent of the S_j's. For example, if

$$x(t) = r \cos(\omega t + \phi),$$

$x(t)$ converges to $c \times S$ where c and S are independent, S is an arcsine law with density defined in (6.15) and $c = \sqrt{x^2(0) + (x'(0)/\omega)^2}$.

Next consider a nonhomogeneous, nonautonomous system of first order differential equations

$$x'(t) = Ax(t) + b(t), \tag{6.16}$$

where A is an n by n matrix, $b : \mathbb{R} \to \mathbb{R}^n$ a continuous function and $x : \mathbb{R} \to \mathbb{R}^n$ a differentiable function with derivative $x'(t)$. The general solution of (6.16) is

$$x(t) = e^{tA} \left\{ x(0) + \int_0^t e^{-As} b(s) ds \right\}.$$

The possible sources of randomness are the values of the parameters that determine the model, that is, A and $b(t)$, and the initial conditions, $x(0)$. If the law determining the behavior of $x(t)$ is fixed, $x(0)$ is the only source of randomness. Assume $x(0)$ is a random vector that satisfies the hypothesis of Theorem 6.2 and that all eigenvalues of A have positive real parts. Denote by P the matrix with columns equal to a basis of generalized eigenvectors associated to the Jordan form of A. Theorem 6.2 implies that

$$d_V\big(x(t)(\mathrm{mod}\,1), U_n\big) \leq \frac{1}{8} W\big(x(0)\big) \|e^{-tA}\|_\infty$$

$$\leq cW\big(x(0)\big) \left\{ \sum_{k=0}^{p} \frac{t^k}{k!} \right\} e^{-(\mathrm{Re}\lambda)t}, \tag{6.17}$$

with $c = \|P\|_\infty \|P^{-1}\|_\infty / 8$, $W\big(x(0)\big) = \sum V_i\big(x(0)\big)$, U_n a distribution uniform on $[0,1]^n$, λ the eigenvalue of A with smallest real part and p its multiplicity. The bound in (6.17) depend on **(a)** how orthogonal the generalized eigenvectors of A are, as measured by the term $\|P\|_\infty \|P^{-1}\|_\infty$ (called P's conditioning number), **(b)** how large the real parts of the eigenvalues of A are, as measured by the term $e^{-(\mathrm{Re}\lambda)t}$, and **(c)** how smooth the density of $x(0)$ is, as quantified by $W\big(x(0)\big)$.

It is interesting to note that if A is diagonalizable, the polynomial term in (6.17) is equal to one and $x(t)(\mathrm{mod}\,1)$ converges to U_n in the variation distance at an exponential rate.

The case where some eigenvalues of A have negative real part can be analyzed using Theorem 6.14. For example, if A has no purely imaginary eigenvalues and there does not exist a non zero integer valued vector orthogonal to all columns of A associated to eigenvalues with positive real part, $x(t)(\mathrm{mod}\,1)$ converges in the weak-star topology to a distribution uniform on $[0,1]^n$.

6.3 Automorphisms of the n-dimensional torus

A continuous automorphism of the n dimensional torus may be viewed as a function

$$T_A : [0,1]^n \to [0,1]^n$$
$$X \to (AX)(\mathrm{mod}\,1), \tag{6.18}$$

where A is an n by n matrix with integer coefficients and determinant equal to ± 1 and X a random vector distributed uniformly on the unit hypercube.

The transformation T_A is invertible and measure-preserving, the latter meaning that the preimage through T_A of any measurable subset B of $[0,1]^n$, $T_A^{-1}(B)$, has the same Lebesgue measure as B.

Automorphisms of the n dimensional torus play an important role in ergodic theory. They are among the best understood measure preserving transformations and some of ergodic theory's main problems have only been solved for transformations which are not significantly more general.

Walters (1982, p.23f) points out that there are two type of problems in measure-theoretic ergodic theory. The first type, called *internal problems*, are concerned with deciding when two measure-preserving transformations are isomorphic. The second type involves studying the kth iterate of a measure preserving transformation T, T^k, for large values of k.

For a continuous automorphism of the n dimensional torus, T_A, define T_A^k inductively as follows:

$$T_A^k(X) = \begin{cases} T_A(X), & k = 1; \\ T_A(T_A^{k-1}(X)), & k = 2, 3, \ldots. \end{cases}$$

Since the determinant of A is equal to ± 1, T_A is invertible and T_A^{-k}, $k = 1, 2, \ldots$ can be defined based on T_A's inverse in a similar way. As the coefficients of A are integers, $T_A^k(X) = (A^k X)(\mathrm{mod}\,1)$, and studying the behavior of $T_A^k(X)$ for large k is equivalent to studying $(A^k X)(\mathrm{mod}\,1)$ as k tends to infinity.

In the following discussion some connections are made between the results of this section and more classical ergodic theory.

The transformation T_A is isomorphic to a Bernoulli-shift if there exists a finite partition $\mathcal{P} = \{B_1, \ldots, B_r\}$ of $[0,1]^n$ such that

a) The sigma-algebra generated by $\{A^k(B_i); k \in \mathbb{Z}, i = 1, \ldots, r\}$ is that of all Borel sets in $[0,1]^n$.

b) The stochastic process $\ldots, V_{-1}(X), V_0(X), V_1(X), \ldots$ with $V_k(X)$ equal to j if the random vector $(A^k X)(\mathrm{mod}\,1)$ belongs to B_j, is a process of independent, identically distributed random variables.

Condition a) ensures that if x and y in $[0,1]^n$ are different then the corresponding sequences $\left(V_k(x)\right)_{k \in \mathbb{Z}}$ and $\left(V_k(y)\right)_{k \in \mathbb{Z}}$ differ at some point and therefore no information is lost by considering the process induced by the partition $\mathcal{P}$.

Condition b) can be interpreted as saying that knowing in which member of the partition $\mathcal{P}$ the process $(A^k X)(\mathrm{mod}\,1)$ is at all times before time k is of no use in predicting future values of the process.

Adler and Weiss (1970) showed that the automorphism of the two dimensional torus induced by the matrix A is isomorphic to a Bernoulli shift if and only if A has no eigenvalue on the unit circle. Katznelson (1971) proved that this is also the case for the n dimensional torus. Extensions to continuous ergodic automorphisms of a compact, metric, abelian group are due to Lind (1977) and Miles and Thomas (1978).

What can be said about a continuous automorphisms of the torus from the point of view of the method of arbitrary functions? Assume the random vector X has a characteristic function that vanishes at infinity (e.g., a density) and A is an n by n matrix with integer coefficients, determinant equal to ± 1 and no eigenvalue on the unit circle. Proposition 6.13 implies that, as k tends to infinity, the infinite dimensional process

$$Y_k = ((A^k X)(\mathrm{mod}\,1), (A^{k+1} X)(\mathrm{mod}\,1), (A^{k+2} X)(\mathrm{mod}\,1), \ \ldots\ ) \qquad (6.19)$$

converges weak-star to a process L on $[0,1]^\infty$ that does not depend on X. The s dimensional marginal distributions of this process are uniform distributions induced by subgroups of $\mathbb{Z}^s$. Due to the assumptions made on A, at least one of its eigenvalues lies outside the unit circle and therefore its one dimensional marginals are uniform on $[0,1]$.

Using the concepts introduced in Chap. 5, the dynamical system that associates to a random vector X the process defined in (6.19) is statistically regular, with respect to the weak-star topology, as k tends to infinity.

While ergodic theory makes the idea of unpredictability of a continuous automorphism of the torus precise by showing that the process Y_k defined in (6.19) is equivalent to a Bernoulli process when X has a distribution uniform on $[0,1]^n$, the method of arbitrary functions shows that for any X with characteristic function vanishing at infinity, Y_k has a non trivial limit that does not depend on the distribution of X.

Remark 1. Due to Remark 3 following Proposition 6.12, the distribution of the limit L cannot be uniform on $[0,1]^\infty$.

Remark 2. Continuous endomorphisms of the n dimensional torus may also be viewed as functions like T_A defined in (6.18), requiring that A be non singular instead of having determinant equal to ± 1. Assume A has at least one eigenvalue outside the unit circle and no eigenvalue on the unit circle. Proposition 6.13 shows that the process Y_k defined in (6.19) has a non trivial limit that does not depend on the distribution of X if the characteristic function vanishes at infinity.

Remark 3. Assume the matrix A defines a continuous automorphism of the two dimensional torus. It is well known that the eigenvalues of A then are irrationals. One of them has absolute value larger than one, while the other one has absolute value less than one. Proposition 6.16 combined with Propositions 2.13 and 2.14 can be used to compute explicit bounds for the rate at which the random vector $(A^k X)(\mathrm{mod}\,1)$ converges to U_2 when the coordinates of X have bounded mean-conditional variation with respect to each other. The example following Proposition 6.16 computes these bounds for the automorphism associated to $\begin{pmatrix} 1 & 1 \\ 1 & 2 \end{pmatrix}$.

References

Adler, R.L., Weiss, B. (1970): Similarity of automorphisms of the torus. Memoirs of the American Mathematical Society, No. 98. Providence, Rhode Island

Anselone, P.M. (1973): In honor of Professor Eberhard Hopf on the occasion of his seventieth birthday. Applicable Analysis **3**, 1–5

Arnold, V.I. (1978): Mathematical Methods of Classical Mechanics. Springer, New York

Barger, V. D., Olsson, M. G. (1973): Classical Mechanics: A Modern Perspective. McGraw-Hill, New York

Bass, T.A. (1985): The Eudaemonic Pie: Or Why Would Anyone Play Roulette Without a Computer in His Shoe? Houghton Mifflin, Boston

Billingsley, P. (1968): Convergence of Probability Measures. Wiley, New York

Billingsley, P. (1986): Probability and Measure (2nd Ed.) Wiley, New York

Borel, E. (1909): Eléments de la Théorie des Probabilités. Hermann, Paris. [English transl.: Prentice Hall, New Jersey, (1965)]

Butzer, P.L., Nessel, R.J. (1971): Fourier Analysis and Approximation. (One–Dimensional Theory, vol. 1.) Academic Press, New York

Collet, P., Eckmann, O.P. (1980): Iterated Maps on the Interval as Dynamical Systems. Birkhäuser, Boston

Copeland, A.H. (1936): A mixture theorem for nonconservative mechanical systems. Bull. Amer. Math. Soc. **42**, 895–900

Cornfeld, I.P., Fomin, S.V. and Sinai, Y.G. (1982): Ergodic Theory. Springer, New York

Diaconis, P., Engel, E. (1986): Comment. Statistical Science. **1**, 171–174

Engel, E. (1987): A Road to Randomness. Ph.D. Thesis, Dep. of Statistics, Stanford

Feller, W. (1971): An Introduction to Probability Theory and Its Applications. (Vol. 2, 2nd Ed.) Wiley, New York

Fréchet, M. (1921): Remarque sur les probabilités continues. Bulletin des Sciences Mathématiques. (2e serie). **45**, 87–88

Garsia, A.M. (1962): Arithmetic properties of Bernoulli convolutions. Trans. Am. Math. Soc. **102**, 409–432

Goldstein, H. (1980): Classical Mechanics. (2nd Ed.) Addison-Wesley, Reading, Massachusetts

Good, I.J. (1986): Some Statistical Applications of Poisson's Work. Statistical Science. **1**, 157–180

Hardy, G.H., Littlewood, J.E. (1922): Some problems of Diophantine approximation: The lattice points of a right–angled triangle (Second Memoir). Abh. Math. Sem. Hamburg. **1**, 212–249

Hewitt, E., Stromberg, K. (1965): Real and Abstract Analysis. Springer, New York

Hoffman, K., Kunze, R. (1971): Linear Algebra. (2nd Ed.) Prentice Hall, New Jersey

Hopf, E. (1934): On causality, statistics and probability. Jour. Math. and Physics, MIT. **13**, 51–102

Hopf, E. (1936): Über die Bedeutung der willkürlichen Funktionen für die Wahrscheinlichkeitstheorie. Jahresber. Deutscher Math. Ver. **46**, 179–195

Hopf, E. (1937): Ein Verteilungsproblem bei dissipativen dynamischen Systemen. Mathematische Annalen. **114**, 161–186

Inoue, H., Kumahora, H., Yoshizawa, Y., Ichimura, M. Miyitake, D. (1983): Random numbers generated by a physical device. Appl. Statist. **32**, 115–120

Kallenberg, O. (1980): Convergence of non-ergodic dynamical systems. Z. Wahrscheinlichkeits-theorie und verw. Gebiete. **53**, 329–351

Katznelson, I. (1971): Ergodic automorphisms of T^n are Bernoulli shifts. Israel J. Math. **10**, 186–195

Keller, J.B. (1986): The probability of heads. Amer. Math. Monthly. **93**, 191–196

Kemperman, J. (1975): Sharp bounds for discrepancies (mod 1) with application to the first digit problem. Unpublished manuscript

Khinchin, A.Y. (1935, 1964): Continued Fractions. (3rd Ed.) University of Chicago Press, Chicago

Kolmogorov, A.N. (1933, 1956): Foundations of the Theory of Probability. Chelsea, New York

Kuipers, L., Niederreiter, H. (1974): Uniform Distribution of Sequences. Wiley, New York

Lehmann, E. (1983): Theory of Point Estimation. Wiley, New York

Lind, D. (1977): The structure of skew products with ergodic group automorphisms. Israel J. Math. **28**, 205–248

Loève, M. (1977): Probability Theory (Vol. 1, 4th Ed.) Springer, New York

Maor, E. (1972): A repertoire of S.H.M. The Physics Teacher.

Miles, G., Thomas, K. (1978): Generalized torus automorphisms are Bernoullian. Studies in Probability and Ergodic Theory. Advances in Math., Supplementary Studies, Vol. 2, 231–249

Moser, J. (1973): Stable and Random Motions in Dynamical Systems. Princeton University Press, New Jersey

Niederreiter, H., Philipp, W. (1973): Berry-Esseen bounds and a theorem of Erdös and Turán on uniform distribution Mod 1. Duke Math. J. **40**, 633–649

Pitman, E.J.G. (1979): Some Basic Theory for Statistical Inference. Chapman and Hall, London

von Plato, J. (1983): The method of arbitrary functions. British J. Phil. Sci. **34**, 34–47

Poincaré, H. (1896): Calcul des Probabilités. Gauthier-Villars, Paris

Poincaré, H. (1905, 1952): Science and Method. Walter Scott and Dover, New York

Rényi, A. (1958): On mixing sequences of sets. Acta Math. Acad. Sci. Hung. **9**, 215–228

Rényi, A., Révész P. (1958): On mixing sequences of random variables. Acta Math. Acad. Sci. Hung. **9**, 389–393

Ripley, B.O. (1987): Stochastic Simulation. Wiley, New York

Roth, K.F. (1955): Rational approximations to algebraic numbers. Mathematika. **2**, 1–20 and 168

Royden, H.L. (1968): Real Analysis. (2nd Ed.) MacMillan, New York

Savage, L.J. (1973): Probability in science: a personalistic account. In P. Suppes, L. Henkin, A. Joja, G.C. Moisil (eds.) Logic, Methodology and Philosophy of Science IV. North-Holland, Amsterdam

Siegel, C.L. (1921): Approximationen algebraischer Zahlen. Math. Zeitschr. **10**, 173–213

Stein, E.M., Weiss, G. (1971): Introduction to Fourier Analysis on Euclidean Spaces. Princeton University Press, Princeton, New Jersey

Ulam, S.M., Neumann, J.v. (1947): On combinations of stochastic and deterministic processes. Bull. Amer. Math. Soc. **53**, 1120.

Vulovic, V.Z., Prange, R.E. (1985): Is the toss of a true coin really random? Preprint, University of Maryland

Walters, P. (1982): An Introduction to Ergodic Theory. Springer, New York

Whittle, P. (1983): Prediction and Regulation (2nd Ed). U. of Minnesota Press, Minnesota

Yue, Z., Zhang, B. (1985): On the sensitive dynamical system and the transition from the apparently deterministic process to the completely random process. Applied Math. and Mech., English Edition. **6**, 193-211

Index